循序渐进学 AI 系列丛书

Java EE 企业级应用开发技术研究

杨树林 著

电子工业出版社
Publishing House of Electronics Industry
北京·BEIJING

内 容 简 介

Java EE 技术是当今流行的 Web 程序开发技术之一。本书结合企业级应用的需要，对 Java EE 的主流技术进行了研究，并结合应用实际讲解了其开发环境、建模方法、架构模式，以及基于主流架构技术的设计方法。全书共 7 章，内容依次为：Java EE 与企业级应用开发、基于 Spring Boot 构建项目、基于 Spring Data JPA 实现数据访问层、基于 MyBatis 实现数据访问层、基于 Spring Security 实现认证和授权、微服务架构与 Spring Cloud、在微服务架构中整合 OAuth2。另外，本书通过一个网上人才中心系统的设计案例，分别介绍了单体架构和微服务架构涉及的核心技术，演示了构建一个安全可靠、稳定高效、易于扩展的应用系统的方法。

本书内容丰富、讲解详尽，适合作为相关研究人员的参考书，也适合作为软件开发人员及其他有关人员的技术参考书。

未经许可，不得以任何方式复制或抄袭本书之部分或全部内容。
版权所有，侵权必究。

图书在版编目（CIP）数据

Java EE 企业级应用开发技术研究 / 杨树林著. —北京：电子工业出版社，2021.1
（循序渐进学 AI 系列丛书）
ISBN 978-7-121-39941-1

Ⅰ．①J… Ⅱ．①杨… Ⅲ．①JAVA 语言－程序设计 Ⅳ．①TP312.8

中国版本图书馆 CIP 数据核字（2020）第 224811 号

责任编辑：徐蔷薇　　　特约编辑：田学清
印　　刷：三河市鑫金马印装有限公司
装　　订：三河市鑫金马印装有限公司
出版发行：电子工业出版社
　　　　　北京市海淀区万寿路 173 信箱　　邮编：100036
开　　本：787×1092　1/16　　印张：18.25　　字数：456 千字
版　　次：2021 年 1 月第 1 版
印　　次：2021 年 1 月第 1 次印刷
定　　价：88.00 元

凡所购买电子工业出版社图书有缺损问题，请向购买书店调换。若书店售缺，请与本社发行部联系，联系及邮购电话：（010）88254888，88258588。
质量投诉请发邮件至 zlts@phei.com.cn，盗版侵权举报请发邮件至 dbqq@phei.com.cn。
本书咨询联系方式：xuqw@phei.com.cn。

前言

随着信息技术的飞速发展和商业竞争的加剧，开发高效、安全的企业级应用系统已经成为各企业提高竞争能力的重要手段。分布式的、高性能的企业级开发平台被越来越多的开发人员使用，Java EE（Java Platform, Enterprise Edition）是这些平台的优秀代表之一。为了缩减开发成本、提高企业软件开发应用的质量，Java EE 为开发者提供了简化的、基于组件的设计方法，用于开发、集成和部署应用程序。Java EE 还提供了多层分布式应用模式，使组件具有重用的能力。在 Java EE 应用开发过程中，框架和模式是很重要的软件重用技术，它们的设计和应用在很大程度上决定了 Java EE 项目的质量。MVC 设计模式分离了数据的控制和数据的表现功能，在实现多层 Web 应用系统中具有明显的优势。Spring MVC 是一个基于 Java EE 平台、实现了 MVC 设计模式的框架，它全面减轻了构建多层 Web 应用的负担，提供了可复用的软件架构。Spring Boot 引入了自动配置的概念，让项目配置变得更容易，进一步简化了 Spring 应用的整个搭建和开发的过程。

近几年来，企业的 IT 环境和 IT 架构也逐渐发生变革，从过去的单体架构（Monolithic 架构）发展为至今广为流行的微服务架构（Microservice 架构）。微服务是一种架构风格，可以给软件应用开发带来很大的便利，但微服务的实施和落地面临很大的挑战，因此需要一套完整的微服务解决方案。Spring Cloud 正是为了适应这种需要而诞生的。它是一个基于 Spring Boot 实现的服务治理工具包，在微服务架构中用于管理和协调服务的微服务。它利用 Spring Boot 的开发便利性巧妙地简化了分布式系统基础设施的开发，如服务发现与注册、配置中心、消息总线、负载均衡、熔断器、数据监控等，都可以使用 Spring Boot 的开发风格实现一键启动和部署。

本书首先分析了企业级应用的特点和常用技术、最新的开发环境、建模方法，以及架构模式；然后重点研究了主流的架构技术 Spring Boot、持久化 ORM 技术 Spring Data JPA 和 MyBatis、安全认证技术 Spring Security、微服务技术 Spring Cloud，以及统一认证技术 OAuth2；同时着重研究了如何结合实际应用进行合理的、快捷的、安全的设计，为实际设计和开发一个企业级应用奠定了基础。本书的主要特点包括：

- 根据市场需求的应用精心选取内容，合理组织内容结构。
- 注意新方法、新技术的引用，突出实用内容。

● 注重方案的选择和设计。

通过系统地学习本书，读者可以将 Java EE 的设计理念快速应用于生产实践中，为开发团队和企业提供坚不可摧的竞争力。

全书共 7 章，内容依次为：Java EE 与企业级应用开发、基于 Spring Boot 构建项目、基于 Spring Data JPA 实现数据访问层、基于 MyBatis 实现数据访问层、基于 Spring Security 实现认证和授权、微服务架构与 Spring Cloud、在微服务架构中整合 OAuth2。

由于时间和作者水平所限，书中难免存在疏漏和不足之处，恳请读者批评指正，以使本书得以改进和完善。

作者

2020 年 8 月于北京

目录

第1章 Java EE 与企业级应用开发 1
1.1 Java EE 概述 1
1.1.1 企业级应用及其特点 1
1.1.2 Java EE 及其常用技术 2
1.2 搭建 Java EE 开发环境 5
1.2.1 安装与配置 JDK 5
1.2.2 安装与使用 MySQL 6
1.2.3 安装集成开发工具 IDEA 8
1.2.4 安装 Tomcat 14
1.3 建模工具与编码规范化 17
1.3.1 UML 建模工具 PlantUML 17
1.3.2 数据库建模工具 Workbench 25
1.3.3 IDEA 数据库管理工具 26
1.3.4 基本编码规范与常用技术 29
1.3.5 Spring Boot 集成 Swagger2 35
1.4 Java EE 项目的分层架构模式 39
1.4.1 分层架构模式概述 39
1.4.2 Java Web 应用中的三层结构 40
1.4.3 结合 MVC 模式的分层结构 41
1.4.4 网上人才中心系统分析与设计 42

第2章 基于 Spring Boot 构建项目 54
2.1 Spring Boot 概述 54
2.1.1 Spring 及 Spring MVC 54
2.1.2 Spring Boot 56
2.2 使用 IDEA 创建 Spring Boot 项目 57
2.2.1 创建 Spring Boot 项目 57
2.2.2 根据项目需要引入其他依赖 60
2.2.3 按分层结构组织程序结构 63
2.2.4 建立分页工具类 65

2.2.5 应用程序基本配置 ... 67
2.3 实体类与接口设计 ... 67
　2.3.1 实体类设计 ... 67
　2.3.2 业务逻辑层接口设计 ... 69
　2.3.3 数据访问层接口设计 ... 69
2.4 数据访问层与业务逻辑层实现 ... 69
　2.4.1 数据访问层实现 ... 69
　2.4.2 对数据访问层进行单元测试 ... 72
　2.4.3 业务逻辑层实现 ... 75
2.5 控制层实现 ... 76
　2.5.1 控制层设计的基本原理 ... 76
　2.5.2 控制类基类设计 ... 78
　2.5.3 实现其他控制类 ... 79
　2.5.4 对控制层进行单元测试 ... 81
2.6 视图层实现 ... 83
　2.6.1 系统首页设计 ... 84
　2.6.2 管理员视图设计 ... 85
　2.6.3 部署运行程序 ... 95

第3章 基于 Spring Data JPA 实现数据访问层 ... 98
3.1 Spring Data JPA 概述 ... 98
　3.1.1 ORM 与 JPA ... 98
　3.1.2 Spring Data JPA ... 99
　3.1.3 Spring Data JPA 接口和类 ... 100
3.2 Spring Boot 与 Spring Data JPA 整合 ... 103
　3.2.1 Spring Data JPA 基本配置 ... 103
　3.2.2 数据源配置优化 ... 104
　3.2.3 基于 Spring Data JPA 实现 Dao 层 ... 107
　3.2.4 Spring Data JPA 扩展 ... 108
3.3 实体对象映射 ... 110
　3.3.1 实体映射基础 ... 110
　3.3.2 实体关系映射 ... 112
　3.3.3 使用逆向工程生成实体类 ... 117
　3.3.4 网上人才中心系统实体类定义 ... 119
3.4 JPA 数据操作方法 ... 123
　3.4.1 使用预定义的方法查询 ... 123
　3.4.2 使用自定义方法查询 ... 126
　3.4.3 查询结果格式 ... 128
　3.4.4 网上人才中心系统数据访问层设计 ... 129

3.4.5 网上人才中心系统业务逻辑层设计 129

第4章 基于MyBatis实现数据访问层 132

4.1 MyBatis 技术概述 132
4.1.1 MyBatis 简介 132
4.1.2 MyBatis 与 Spring Data JPA 比较 132
4.1.3 MyBatis 核心类及工作原理 133
4.1.4 映射器与 Mapper 实例 135

4.2 Spring Boot 与 MyBatis 整合 137
4.2.1 MyBatis 基本配置 137
4.2.2 基于 MyBatis 实现 Dao 层 137
4.2.3 MyBatis 映射器配置 140

4.3 基于 MyBatis Generator 的逆向工程 145
4.3.1 MyBatis Generator 基础 145
4.3.2 MyBatis Generator 扩展 149
4.3.3 使用自动生成的代码操作数据库 151

4.4 基于 MyBatis-Plus 的逆向工程 152
4.4.1 MyBatis-Plus 基础 152
4.4.2 MyBatis-Plus 扩展 156
4.4.3 基于 MyBatis-Plus 的数据操作 159

第5章 基于Spring Security实现认证和授权 162

5.1 Spring Security 概述 162
5.1.1 Spring Security 简介 162
5.1.2 Spring Security 原理 162
5.1.3 Spring Security 配置基础 164

5.2 网上人才中心系统权限体系设计与开发 169
5.2.1 权限相关数据结构及实体类设计 169
5.2.2 权限相关数据访问层设计 174
5.2.3 权限相关业务逻辑层设计 178
5.2.4 权限相关控制层设计 180
5.2.5 权限相关视图层设计 187

5.3 权限相关组件设计及其配置设计 193
5.3.1 权限相关组件设计 193
5.3.2 验证码实现相关设计 197
5.3.3 权限相关配置设计 200

第6章 微服务架构与Spring Cloud 203

6.1 微服务架构概述 203
6.1.1 单体架构与微服务架构 203

  6.1.2 Spring Cloud 概述 206
  6.1.3 Spring Cloud 重要组件介绍 207
 6.2 网上人才中心系统微服务工程设计 208
  6.2.1 微服务设计基础 208
  6.2.2 微服务项目结构 209
  6.2.3 创建微服务项目 210
  6.2.4 创建模块 215
 6.3 基础微服务项目设计 216
  6.3.1 创建服务注册中心 216
  6.3.2 创建配置管理中心 217
  6.3.3 创建微服务网关 221
 6.4 REST API 微服务设计 223
  6.4.1 领域业务设计 223
  6.4.2 查询对象设计 227
  6.4.3 REST API 应用设计 228
  6.4.4 RESTful 的 HTTP 接口设计 236
 6.5 视图微服务设计 239
  6.5.1 Thymeleaf 技术 239
  6.5.2 Web UI 微服务设计 241
  6.5.3 统一入口微服务设计 250

第 7 章　在微服务架构中整合 OAuth2 259

 7.1 基于 OAuth2 实现 SSO 的原理 259
  7.1.1 OAuth2 基本原理 259
  7.1.2 JWT 概述 260
  7.1.3 在微服务架构中实现 SSO 261
 7.2 OAuth2 授权服务器模块设计 262
  7.2.1 OAuth2 授权服务器模块 262
  7.2.2 对授权服务器进行配置 264
  7.2.3 登录管理及安全配置 267
  7.2.4 控制器和用户登录界面设计 270
 7.3 实现微服务应用访问控制 272
  7.3.1 对网关 Zuul 进行配置 272
  7.3.2 创建安全模块 273
  7.3.3 配置微服务应用 280

参考文献 281

第 1 章　Java EE 与企业级应用开发

本章主要介绍企业级应用及其特点、Java EE 及其常用技术、搭建 Java EE 开发环境、建模工具与编码规范化，以及 Java EE 项目的分层架构模式。

1.1　Java EE 概述

1.1.1　企业级应用及其特点

1．企业级应用概念

企业级应用是指为商业组织、大中型企业而创建的解决方案及应用程序。这些大中型的企业级应用具有用户数量多、数据量大、事务密集等特点，往往需要满足未来业务需要的变化，易于升级和维护。

一个好的企业级应用体系结构，通常来自优秀的解决方案，同时从应用程序的设计开始就需要考虑其体系结构的合理性、灵活性、健壮性，从而既能满足企业级应用的复杂需求也能为今后系统的调整和升级留有余地。这样可以延长整个应用的生命周期，增强用户在多变的商业社会中的适应性，减少系统维护的开销和难度，从而给用户带来最大的利益。

企业级应用通常具有以下特点。

（1）数据持久化（Persistent Data）。企业级应用需要持久地保存数据，通常需要保存很多年。在这段时间里，使用数据的程序会经常发生改变。有时企业为了处理一个业务，安装了一个全新的应用系统，那么这些数据也必须被移植到新的应用系统上。

（2）数据的并发访问。多用户并发地存取数据是企业级应用的常见情况。对于基于 Internet 的 Web 系统来说，使用者数量的递增速率是几何级的。因此确保大量使用者都能从系统中正常地访问数据就是一个非常重要的问题。即使没有那么多用户，也要保证两个人不会在同一时刻对同一个数据进行存取。然而用户数量过多所带来的沉重压力，通过事务管理工具仅仅只能解决一部分。

（3）大量的用户图形界面。为了应付日益庞大的数据量，大量的 UI 界面被投入使用，所以即使出现成百上千个截然不同的界面也并不奇怪。普通用户与专业用户的习惯差异很大，他们很少有技术层面的专长。为了满足不同的需求，数据的表现形式是千差万别的。

（4）需要和其他企业级应用集成。企业级应用并不是"信息孤岛"，它们经常需要和其他企业级应用集成在一起。这些企业级应用通常是在不同的时期，采用不同的技术创建的，甚至协作的机制也各不相同。企业会尽力将不同的企业级应用通过一个通用的通信技术集成起来，但是即使这样也很难圆满地完成任务，所以企业会同时使用几套不同的集成方案。

（5）数据概念不统一。即使统一了集成的技术，也经常会碰到千差万别的业务处理方式和不统一的数据概念等问题。企业的一个部门可能会认为，顾客是一个与公司拥有正式协议的人；另一个部门可能会认为，那些曾经与公司签订过合同人也是顾客，即使现在已经解除了；还有一个部门可能会认为，产品销售面向的人是顾客，而服务销售面向的人不是。一开始，我们可能会感觉这很简单，很好解决，但是当成百上千条记录在每个领域都有截然不同的意思时，即使公司里有人能够区分不同领域的不同意思，我们也将面临严峻的挑战。结果数据会被经常读取，并按照各种各样的不同的语法或语义格式记录下来。

（6）复杂的业务逻辑。业务逻辑是由企业根据自身需要所制定的业务规则决定的。有时规则会很随意，看似没有任何的逻辑。这是因为企业往往有自己特定的需求，以及特殊的情况。这些层出不穷的特例导致了业务的复杂性、无逻辑性，使得商业软件的开发十分困难。

2. 企业级应用的要求

现代企业级应用是以服务器为中心，通过网络把服务器和分散的用户联系在一起的应用。一般来说，现代企业级应用应当具有以下需求。

- 并发支持：同时收到大量服务请求，需要进行快速响应。
- 事务支持：支持事务完整性，对于多个系统，需要支持分布式事务。
- 交互支持：支持系统与系统之间进行交互，人与系统之间进行交互。
- 集群支持：提供系统可用性和可伸缩性，便于随业务的需求而扩展。
- 安全支持：受保护的资源获得安全保护。
- 分布式支持：支持查找和调用分布式服务。
- Web 支持：基于 Internet 或无线网络。

1.1.2 Java EE 及其常用技术

Java 企业级应用开发涉及的技术结构包括三大部分：一部分是分布式开发；一部分是业务组件开发；一部分是资源整合开发（消息传递和交互）。无论是采用经典的框架还是采用 Spring 等开源框架，Java 企业级应用开发的主要任务都是围绕业务需求展开的。由于 Java 技术方案在整个互联网领域有着广泛的应用基础，所以 Java EE 已经成为主流技术。

1. 什么是 Java EE

Java EE（Java Platform，Enterprise Edition）是 Sun 公司推出的企业级应用程序版本。该版本以前被称为 J2EE，能够帮助我们开发和部署可移植、健壮、可伸缩且安全的服务器端 Java 应用程序。Java EE 是在 Java SE 的基础上构建的，它提供了 Web 服务、组件模型、管理和通信 API，可以用来实现企业级的面向服务体系结构（Service-Oriented Architecture，SOA）和 Web 2.0 应用程序。

Java EE 是早期 Java 技术体系中的重要一环，主要用于解决 Java 在企业级应用开发中遇到的性能问题、安全问题及众多资源整合的问题。但是由于 Java EE 自身所占资源空间过大，后期逐渐被 Spring 等轻量级框架所取代。目前，Java EE 已经交由 Eclipse 基金会管理，并改名为 Jakarta EE 了。所以目前所说的 Java EE 开发通常是指采用 Java 进行企业级应用开发，而不仅指 Java EE 技术本身了。

2. Java EE 常用技术

狭义的 Java EE 是 Sun 公司为企业级应用推出的标准平台，用来开发 B/S 架构软件，可以说是一个框架，也可以说是一种规范。它由一整套服务（Services）、应用程序接口（APIs）和协议构成，对开发基于 Web 的多层应用提供了功能支持。广义的 Java EE 包含各种框架，其中最重要的就是 Spring。Spring 在诞生之初是为了改进 Java EE 开发的体验，后来逐渐成为 Java Web 应用开发的实际标准。我们这里采用广义的 Java EE 概念，并重点介绍新兴的架构技术。

早期的一些核心技术主要包括以下几种。

- Servlet：Servlet 是 Java 平台上的 CGI 技术。Servlet 在服务器端运行，可以动态地生成 Web 页面。与传统的 CGI 和许多其他类似 CGI 的技术相比，Java Servlet 具有更高的效率且更容易使用。对于 Servlet，重复的请求不会导致一个程序的多次转载，它是依靠线程的方式来支持并发访问的。

- JSP：JSP（Java Server Page）是一种实现普通静态 HTML 和动态页面输出混合编码的技术。从这一点来看，JSP 类似于 Microsoft ASP、PHP 等技术。借助了形式上的内容和外观表现的分离，Web 页面制作的任务可以比较方便地分配给页面设计人员和程序员，并方便地通过 JSP 来合成。在运行时，JSP 会被先转换成 Servlet，并以 Servlet 的形态编译运行，因此它的效率和功能与 Servlet 相比没有差别，都具有很高的效率。

- JDBC：JDBC（Java Database Connectivity，数据库访问接口）使数据库开发人员能够用标准 Java API 编写数据库应用程序。JDBC API 主要用来连接数据库和直接调用 SQL 命令执行各种 SQL 语句。利用 JDBC API 可以执行一般的 SQL 语句、动态 SQL 语句及带 IN 和 OUT 参数的存储过程。Java 中的 JDBC 相当于 Microsoft 平台中的 ODBC（Open Database Connectivity）。

- EJB：EJB（Enterprise JavaBeans）定义了一组可重用的组件，开发人员可以利用这些组件，像搭积木一样建立分布式应用。在装配组件时，这些组件需要被配置到 EJB 服务器（一般的 WebLogic、WebSphere 等 J2EE 应用服务器都是 EJB 服务器）中。EJB 服务器作为容器和低层平台的桥梁管理着 EJB 容器，并向该容器提供访问系统服务的能力。所有的 EJB 实例都运行在 EJB 容器中。EJB 容器提供了系统级的服务，可以控制 EJB 的生命周期。EJB 容器帮助开发人员管理了安全、远程连接、生命周期及事务等技术环节，简化了商业逻辑的开发。

- JSF：JSF（Java Server Faces）是 Java Web 应用程序的一个用户界面框架。设计 JSF 的目的在于，极大地缓解在 Java 应用服务器上运行的应用程序的编写和维护的压力，并将这些应用程序的 UI 重新呈现给目标客户端。

- JMS：JMS（Java Message Service，Java 消息服务）是一组 Java 应用接口，用于提供创建、发送、接收、读取消息的服务。JMS API 定义了一组公共的应用程序接口和相关语法，使得 Java 应用能够和各种消息中间件进行通信。使用 JMS 能够最大限度地提升消息应用的可移植性。JMS 既支持点对点的消息通信，也支持发布/订阅式的消息通信。

- JNDI：由于 Java EE 应用程序组件一般分布在不同的机器上，因此需要采用一种机制，以便组件使用者查找和引用组件及资源。在 Java EE 体系中，使用 JNDI（Java Naming and Directory Interface，Java 命名和目录接口）定位各种对象，这些对象包括 EJB、JDBC 数据源及消息连接等。JNDI API 为应用程序提供了一个统一的接口来完成标准的目录操作，如通过对象属性来查找和定位该对象。由于 JNDI 是独立于目录协议的，应用还可以使用 JNDI 访问各种特定的目录服务，如 LDAP、NDS 和 DNS 等。

新兴的架构技术主要包括以下几种。

- Hibernate：Hibernate 是一个面向 Java 环境的 ORM（Object Relation Mapping，对象/关系数据库映射）工具。它对 JDBC API 进行了封装，负责 Java 对象的持久化，在分层的软件架构中位于持久化层，封装了所有数据访问细节，使业务逻辑层可以专注于实现业务逻辑。

- Hibernate JPA：JPA（Java Persistence API，Java 持久化应用程序接口）可以通过标注或 XML 描述对象和关系数据库之间的映射关系，并将实体对象持久化到数据库中。它是 Sun 公司提出的一种规范，提供了操作实体对象，执行数据库查询的统一规范。而 Hibernate JPA 是它的一种实现，应用十分广泛。

- MyBatis：MyBatis 是 Apache 的一个开源项目 iBatis。2010 年，这个项目由 Apache Software Foundation 迁移到了 Google Code，并且改名为 MyBatis。2013 年 11 月，这个项目又迁移到了 GitHub。MyBatis 是一个持久层框架，也属于 ORM 映射，是一个半自动化的，并且支持普通 SQL 查询、存储过程和高级映射的优秀持久层框架。MyBatis 避免了几乎所有的 JDBC 代码编写、参数的手工设置及结果集的检索。MyBatis 使用简单的 XML 或标注来配置和映射原始信息，将接口和 Java 的 POJOs（Plain Ordinary Java Objects，普通的 Java 对象）映射成数据库中的记录。

- Spring：Spring 是一个开源框架，是为了解决企业级应用程序开发的复杂性而创建的。Spring 致力于提供一个以统一、高效的方式构造整个应用，并且可以将单层框架以最佳的组合融合在一起的连贯体系。它基于依赖注入和面向方面技术，大大地降低了应用开发的难度与复杂度，很好地解决了模块之间的耦合性，提高了开发的速度，为企业级应用提供了一个轻量级的解决方案。

- Spring MVC：Spring MVC 属于 Spring 框架的后续产品，提供了基于 MVC（Model-View-Controller）构建 Web 应用程序的全功能模块。在使用 Spring 可插入的 MVC 架构后，可以在使用 Spring 进行 Web 开发时选择使用 Spring 的 Spring MVC 框架或集成其他 MVC 开发框架，如 Struts 等。

- Spring Boot：Spring Boot 是由 Pivotal 团队提供的全新框架，其设计目的是简化 Spring 应用的初始搭建及开发过程。该框架使用了特定的方式进行配置，使得开发人员不再需要定义样板化的配置。通过这种方式，Spring Boot 致力于在蓬勃发展的快速应用开发领域（Rapid Application Development）成为领导者。

- Spring Cloud：Spring Cloud 是一个基于 Spring Boot 实现的服务治理工具包，在微服务架构中用于管理和协调服务的微服务。Spring Cloud 会将一个单体项目拆分为多个微服务，每个微服务都可以进行独立技术选型、独立开发、独立部署、独立运维，并且多个服务可以相互协调、相互配合，最终完成用户的价值。Spring Cloud 是一系列框架的有序集合。它利用 Spring Boot 的开发便利性巧妙地简化了分布式系统基础设施的开发，如服务发现注册、配置中心、消息总线、负载均衡、熔断器、数据监控等都可以使用 Spring Boot 的开发风格实现一键启动和部署。

1.2 搭建 Java EE 开发环境

1.2.1 安装与配置 JDK

1. JDK 简介

JDK（Java Development Kit）是 Sun 公司提供的基础 Java 语言开发工具软件包。其中包含 Java 语言的编译工具、运行工具及类库。其目录结构如下所述。

- bin 目录：包含编译器、解释器和一些工具。
- lib 目录：包含类库文件。
- demo 目录：包含各种演示案例。
- include 目录：包含 C 语言头文件，支持 Java 本地接口与 Java 虚拟机调试程序接口的本地编程技术。
- jre 目录：包含 Java 虚拟机、运行时的类包和 Java 应用启动器。
- sample 目录：包含帮助学习者学习的 Java 例子。
- src.zip：源码压缩文件。

在 bin 目录下包括以下常用工具。

- javac.exe：Java 语言编译器，输出结果为 Java 字节码。
- java.exe：Java 字节码解释器。
- javadoc.exe：帮助文档生成器。
- jar.exe：打包工具。
- appletviewer.exe：小应用程序浏览工具，用于测试并运行 Applet 小程序。

2. JDK 下载与安装

JDK 是一个开源、免费的工具。我们可以登录 Oracle 公司的官方网站下载 JDK 最新版本，网址为 http://www.oracle.com/java/technologies/javase-downloads.html。本书使用的 JDK 版本是 Java SE Development Kit 13.0.2。在下载完成后可得到 jdk-13.0.2_windows-x64_bin.exe 文件。双击该文件，即可开始安装 JDK。在安装过程中，可以选择安装路径和安装组件，如果没有特殊要求，保持默认设置即可。默认的安装路径是 C:\Program Files\Java\jdk13.0.2。然后，设置环境变量如下：

JAVA_HOME=<JSEDK 安装目录>
CLASSPATH=.;%JAVA_HOME%\lib;%JAVA_HOME%\lib\tools.jar
Path=<原 Path>;%JAVA_HOME%\bin;%JAVA_HOME%\jre\bin

1.2.2 安装与使用 MySQL

MySQL 是一个关系数据库管理系统，由瑞典 MySQL AB 公司开发，目前是 Oracle 旗下产品。MySQL 是流行的关系数据库管理系统之一，由于其体积小、速度快、总体拥有成本低，并且开放源码，因此许多中小型网站为了降低网站总体拥有成本而选择了 MySQL 作为网站后台数据库。

MySQL 的安装方法如下所述。

（1）登录 MySQL 的官方网站并下载 MySQL 安装包，网址为 https://dev.mysql.com/downloads/mysql/。MySQL 安装包下载界面如图 1-1 所示。

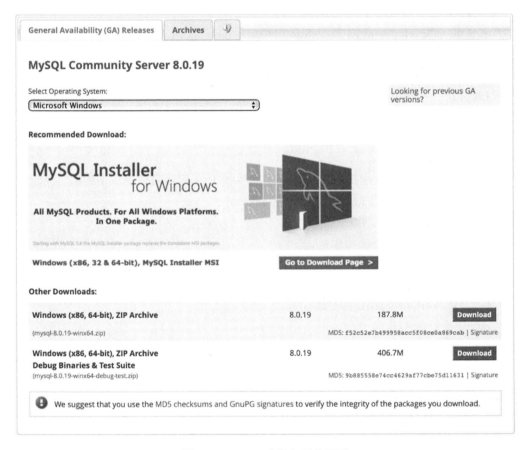

图 1-1　MySQL 安装包下载界面

选择【Windows(x84, 64-bit),ZIP Archive】并单击【Download】按钮，会出现如图 1-2 所示的开始下载界面，单击【No thanks, just start my download】即可直接下载。

图 1-2 开始下载界面

（2）在下载完毕后，将安装包解压到本地。例如，将安装包解压缩到 D:\mysql-8.0.19-winx64 目录下。

（3）添加 my.ini 文件。在上述目录中新建一个 my.ini 文件，内容如下：

```
[mysql]
#设置 MySQL 客户端默认字符集
default-character-set=utf8
[mysqld]
#设置 3306 端口
port = 3306
#设置 MySQL 的安装目录
basedir=D:\mysql-8.0.19-winx64
#设置 MySQL 数据库的数据存放目录
datadir=D:\mysql-8.0.19-winx64\data
#允许最大连接数
max_connections=20
#服务端使用的字符集采用 utf8
character-set-server=utf8
#在创建新表时将使用的默认存储引擎
default-storage-engine=INNODB
#设置协议认证方式（重点）
default_authentication_plugin=mysql_native_password
```

（4）生成 data 文件夹。打开 Windows 命令提示符窗口，进入 MySQL 安装目录的 bin 文件夹，执行如下命令：

```
mysqld --initialize-insecure
```

（5）启动服务。执行如下命令：

net start mysql

（6）登录。执行如下命令：

mysql -u root -p （这里键入回车符，因为之前并未设置密码，密码是空即可）

（7）设置 root 密码。执行如下命令：

update mysql.user set authentication_string="12345" where user="root";

当出现【Query OK】时，证明已经修改成功，但是需要执行如下命令才能保存执行结果：

flush privileges;　　　#如果不执行，则还是之前的密码

（8）配置环境变量。为了方便使用，可以配置环境变量如下：

MYSQL_HOME=D:\mysql-8.0.19-winx64\
path=%MYSQL_HOME%\bin

1.2.3　安装集成开发工具 IDEA

IDEA 全称为 IntelliJ IDEA，是 Java 编程语言开发的集成环境。IDEA 在业界被公认为最好的 Java 开发工具，尤其是在智能代码助手、代码自动提示、重构、J2EE 支持、各类版本工具（如 Git、SVN 等）、JUnit、CVS 整合、代码分析、创新的 GUI 设计等方面的功能可以说是出类拔萃的。IDEA 是 JetBrains 公司的产品。它的旗舰版本支持 HTML、CSS、PHP、MySQL、Python 等语言，而免费版只支持 Python 等少数语言。

IDEA 的安装方法如下所述。

（1）登录 JetBrains 官网的集成开发工具首页（https://www.jetbrains.com/products.html#type=ide），选择【IntelliJ IDEA】，如图 1-3 所示。

图 1-3　选择【IntelliJ IDEA】

（2）在下载界面中选择 Ultimate 版本进行下载，如图 1-4 所示。这里以 Windows 版本为例，在下载完成后可得到 IdeaIU 2019.3.1.exe 文件。

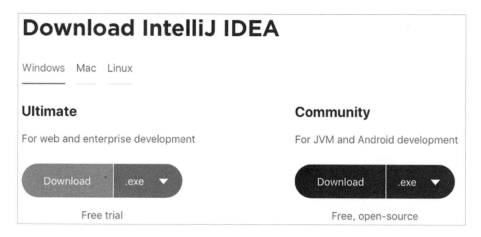

图 1-4 选择 Ultimate 版本

（3）双击 IdeaIU 2019.3.1.exe 文件，开始安装 IDEA，出现欢迎界面，如图 1-5 所示，然后单击【Next】按钮。

图 1-5 欢迎界面

（4）在如图 1-6 所示的界面中选择安装位置，这里保持默认设置，单击【Next】按钮。

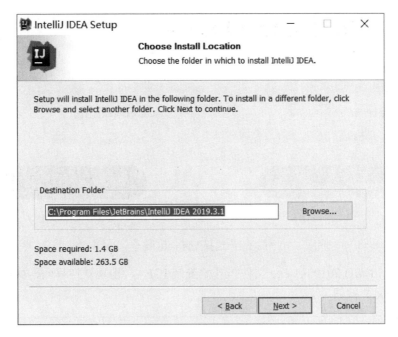

图 1-6　选择安装位置

（5）在如图 1-7 所示的界面中设置选项，勾选【64-bit launcher】和【.java】复选框，单击【Next】按钮。

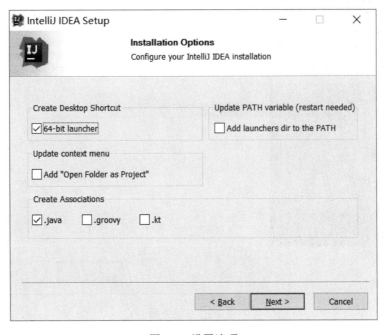

图 1-7　设置选项

（6）在如图 1-8 所示的界面中选择开始菜单文件夹，这里保持默认设置，单击【Install】按钮。

图 1-8　选择开始菜单文件夹

(7) 进入安装过程, 如图 1-9 所示。

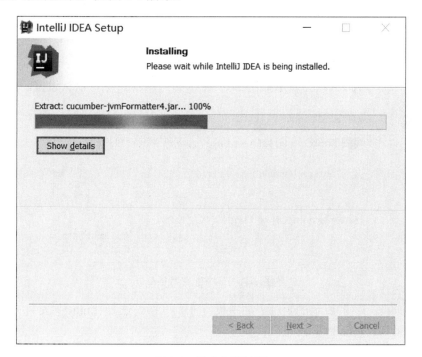

图 1-9　进入安装过程

(8) 在安装完成后,出现如图 1-10 所示的界面,单击【Finish】按钮。

图 1-10　安装完成界面

(9) 初次启动 IDEA,会显示如图 1-11 所示的对话框,保持默认设置,单击【OK】按钮。

图 1-11　环节配置对话框

(10) 在如图 1-12 所示的定制界面中设置 UI 风格。选择 Light 风格,并单击【Skip Remaining and Set Defaults】按钮。

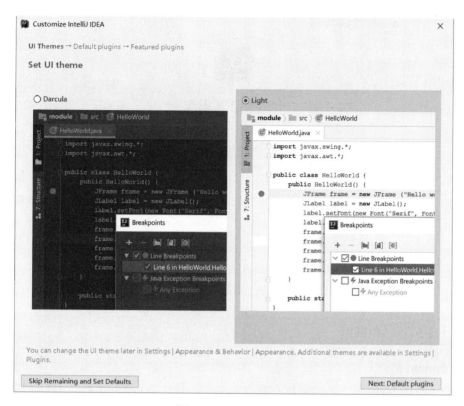

图 1-12　设置 UI 风格

（11）在如图 1-13 所示的界面中激活 IDEA。可通过 3 种方式激活，这里使用了 JB Account 的方式激活。

图 1-13　激活 IDEA

（12）在运行 IDEA 后，进入如图 1-14 所示的运行起始界面，可单击【Create New Project】按钮来创建新项目。

图 1-14　运行起始界面

1.2.4　安装 Tomcat

1. Tomcat 简介

Tomcat 是在 Sun 公司的 JSWDK（JavaServer Web Development Kit，Java 服务器 Web 开发工具）基础上发展起来的一个优秀的 Servlet/JSP 容器。它是 Apache-Jakarta 软件组织的一个子项目，不但支持运行 Servlet 和 JSP，而且具备作为商业 Java Web 应用容器的特征。

作为一个开放源码的软件，Tomcat 得到了源码志愿者的广泛支持。它可以和目前大部分的主流 HTTP 服务器（如 IIS 和 Apache 服务器）一起工作，并且运行稳定、可靠、效率较高。

除了能够运行 Servlet 和 JSP，Tomcat 还提供了作为 Web 服务器的一些特有功能，如 Tomcat 管理和控制平台、安全域管理和 Tomcat 阀等。Tomcat 已成为目前开发企业 Java Web 应用的最佳选择之一。

Tomcat 的具体下载地址为 http://tomcat.apache.org，在下载完成后得到的文件是 apache-tomcat-9.0.31.exe（截至本书编写完成时的最新版本）。

2. Tomcat 的安装步骤

（1）双击 apache-tomcat-9.0.31.exe 文件，打开如图 1-15 所示的 Tomcat 安装向导，单击【Next】按钮。

（2）在如图 1-16 所示的界面中查看许可证协议，并单击【I Agree】按钮。

（3）在如图 1-17 所示的界面中选择需要安装的组件，这里保持默认设置，单击【Next】按钮。

（4）在如图 1-18 所示的界面中进行 Tomcat 基本配置，可设置端口号，管理登录用户名、密码及角色，这里保持默认设置，单击【Next】按钮。

第 1 章　Java EE 与企业级应用开发

图 1-15　Tomcat 安装向导

图 1-16　查看许可证协议

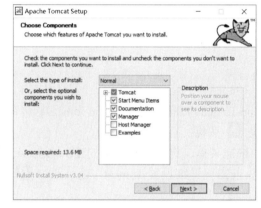

图 1-17　选择需要安装的组件

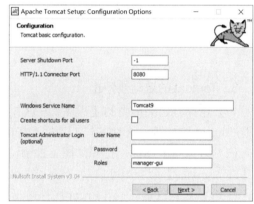

图 1-18　进行 Tomcat 基本配置

（5）在如图 1-19 所示的界面中选择 Java 虚拟机路径，这里采用安装程序自动搜索的虚拟机路径，单击【Next】按钮。

（6）在如图 1-20 所示的界面中选择 Tomcat 安装文件夹，即 Tomcat 安装位置，这里保持默认设置，单击【Next】按钮。

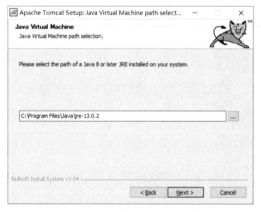

图 1-19　选择 Java 虚拟机路径

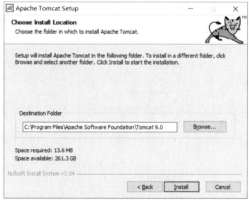

图 1-20　选择 Tomcat 安装文件夹

（7）在如图 1-21 所示的界面中完成安装，可选择是否运行 Tomcat 或显示帮助，这里保持默认设置，单击【Finish】按钮。

图 1-21　完成安装

3．Tomcat 的启动或停止

（1）选择开始菜单中的【程序】→【Apache Tomcat 9.0】→【Monitor Tomcat】命令，打开 Tomcat 监控器，这时在其工具栏上将出现 图标。

（2）右击 图标，打开如图 1-22 所示的快捷菜单，选择该菜单中的【Start service】或【Stop service】命令即可启动或停止 Tomcat。若需要详细设置 Tomcat，则可双击 图标，打开如图 1-23 所示的 Tomcat 属性对话框。

图 1-22　快捷菜单　　　　　　　　图 1-23　Tomcat 属性对话框

1.3 建模工具与编码规范化

1.3.1 UML 建模工具 PlantUML

PlantUML 是一个开源工具，允许用户使用纯文本语言创建 UML 图。PlantUML 的语言是特定于域的语言的示例。它使用 Graphviz 软件来布置图表。PlantUML 具有比较详细的各类语言的 Guide 文档。PlantUML 实现了很多适配，对常用的编译器 Eclipse、IDEA 等都提供了对应的插件，同时 PlantUML 与 Maven 和 jQuery 都进行了集成，还提供了 war 包形式，可以在本地的 Java EE 容器（如 Tomcat）中运行。

1．PlantUML 的使用方式

PlantUML 有如下几种使用方式。

（1）通过网站使用 PlantUML 绘制对应的 UML 图。PlantUML 官网提供了免费的在线编辑服务，可以通过在输入区输入 UML 语言来生成对应的 UML 图，网址为 http://www.plantuml.com/plantuml/uml，如图 1-24 所示。

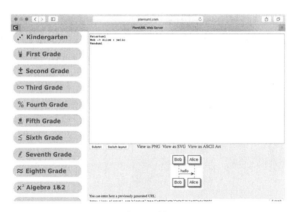

图 1-24　通过网站使用 PlantUML

（2）通过本地的 Java EE 容器运行 plantuml.war 文件。下载 Java J2EE WAR File（plantuml.war），并将其放在本地的 Tomcat 的 webapps 目录下，在启动后访问 http://localhost:8080/plantuml，就可以看到如图 1-25 所示的界面，可以在输入区编写 PlantUML 语言的代码。plantuml.war 文件的下载地址为 https://sourceforge.net/projects/plantuml/files/plantuml.war/download。

（3）在 IDEA 中使用 PlantUML Integration 插件。在使用之前需要安装 PlantUML Integration 插件，然后就可以方便地书写 UML 语言，并生成对应的预览图片。我们可以复制或保存生成的图片到指定的目录，并重启 IntelliJ IDEA。除了安装 PlantUML Integration 插件，还需要下载并安装 Graphviz 到 Windows 上，下载地址为 https://graphviz.gitlab.io/_pages/Download/windows/graphviz-2.38.msi。

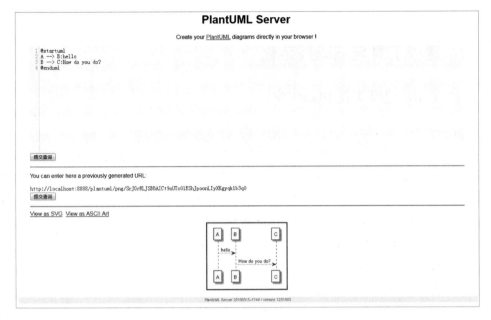

图 1-25　通过 Java EE 容器运行 plantuml.war 文件

2. 在 IDEA 中使用 PlantUML 进行建模

在 IDEA 中绘制 UML 图，可以通过选择【File】→【New】→【PlantUML File】命令来运行 PlantUML，并且在弹出的对话框中选择要创建的 UML 图的类型，输入文件名，如图 1-26 所示。

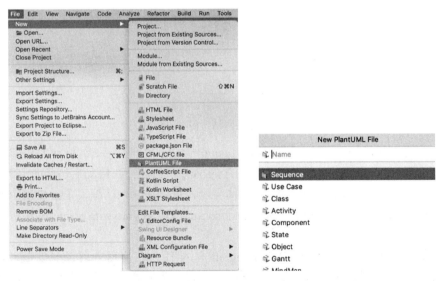

图 1-26　在 IDEA 中绘制 UML 图

这里主要介绍用例图和类图的建模方法。对于其他建模方法，可参见官方网站（http://plantuml.com/zh/）的介绍。

1）用例图

用例图（Use-Case Diagram）是用于描述系统功能的视图，其主要元素是用例和参与者，可以帮助开发团队以一种可视化的方式理解系统的功能需求，并向客户形象化地表述项目的功能。使用 PlantUML 创建用例图的说明如表 1-1 所示。

表 1-1　使用 PlantUML 创建用例图的说明

要　素	描　述	
用例	用例通常使用圆括号括起来。可以使用关键字 usecase 定义用例。还可以使用关键字 as 给用例定义一个别名，这个别名可以在以后定义关系时使用，但有别名就要使用别名。在使用关键字 usecase 定义用例时，可以不使用圆括号，但在含有特殊字符时，必须使用圆括号	
	@startuml (First usecase) (Another usecase) as UC2 usecase UC3 usecase (Last\nusecase) as UC4 @enduml	
角色	角色通常使用两个冒号包裹起来。可以使用关键字 actor 定义角色。还可以使用关键字 as 给角色定义一个别名，这个别名可以在以后定义关系时使用。角色的定义是可选的	
	@startuml :First Actor: :Another\nactor: as Men2 actor Men3 actor :Last actor: as Men4 @enduml	
用例描述	如果想定义跨越多行的用例描述，则可以使用双引号将其引起来，还可以使用分隔符"：""--"".." "==" "__"隔开，并且可以在分隔符中间放置标题	
	@startuml usecase UC1 as "You can use several lines to define your usecase. You can also use separators. -- Several separators are possible. == And you can add titles: ..Conclusion.. This allows large description." @enduml	

续表

要素	描述	
基础示例	使用箭头-->连接角色和用例。横杠-越多，箭头越长。通过在箭头定义的后面添加一个冒号及文字的方式来添加标签。User 不需要定义，直接将其当作一个角色使用即可	
基础示例	@startuml User -> (Start) User --> (Use the application) : A small label :Main Admin: ---> (Use the application) : This is\nyet another\nlabel @enduml	
继承	一个角色或用例继承于另一个角色或用例	
继承	@startuml :Main Admin: as Admin (Use the application) as (Use) User <|-- Admin (Start) <|-- (Use) @enduml	
使用注释	可以使用 note left of、note right of、note top of、note bottom of 等关键字给一个对象添加注释。还可以通过关键字 note 来定义注释，然后使用..连接其他对象	
使用注释	@startuml :Main Admin: as Admin (Use the application) as (Use) User -> (Start) User --> (Use) Admin ---> (Use) note right of Admin : This is an example. note right of (Use) 　A note can also 　be on several lines end note note "This note is connected\nto several objects." as N2 (Start) .. N2 N2 .. (Use) @enduml	

续表

要素	描述	
改变箭头方向	默认连接是竖直方向的,用--表示。可以使用一个横杠或点来表示水平连接。也可以通过翻转箭头来改变箭头方向	

	`@startuml`
	`:user: --> (Use case 1)`
	`:user: -> (Use case 2)`
	`(Use case 3) <.. :user:`
	`(Use case 4) <- :user:`
	`@enduml`
		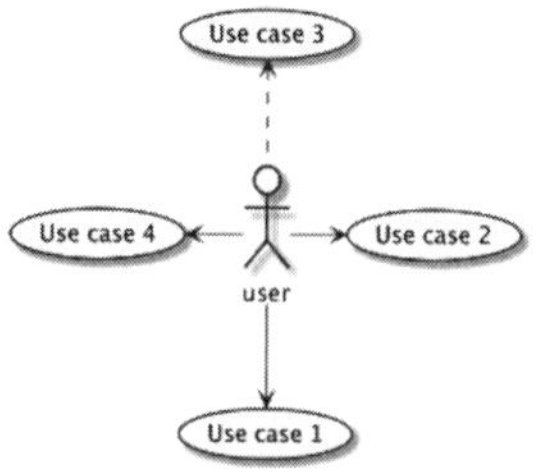

	还可以通过给箭头添加 left、right、up 或 down 等关键字来改变箭头方向	

	`@startuml`
	`:user: -left-> (dummyLeft)`
	`:user: -right-> (dummyRight)`
	`:user: -up-> (dummyUp)`
	`:user: -down-> (dummyDown)`
	`@enduml`
		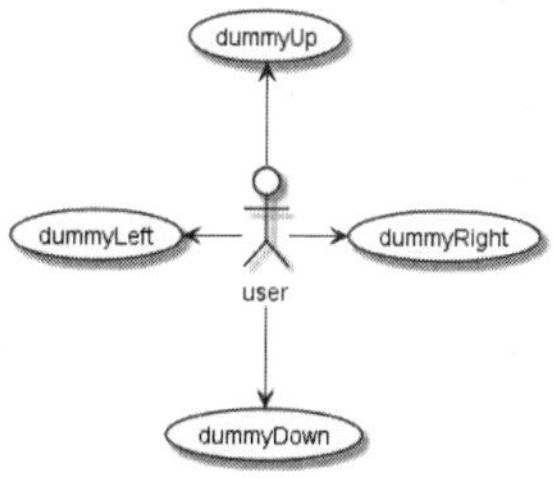

分割图示	使用关键字 newpage 将图示分割为多个页面	

	`@startuml`
	`:actor1: --> (Usecase1)`
	`newpage`
	`:actor2: --> (Usecase2)`
	`@enduml`
		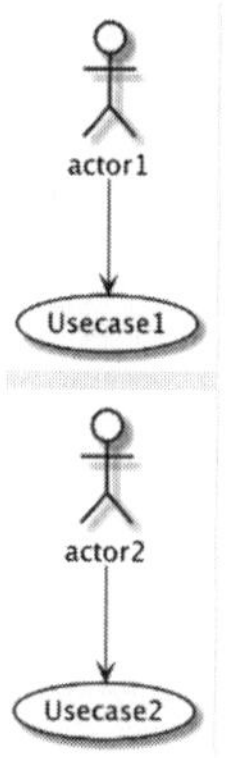

改变图示方向	默认从上向下(top to bottom direction)构建图示。可以使用 left to right direction 命令改变图示方向为从左向右	

	`@startuml`
	`left to right direction`
	`user1 --> (Usecase 1)`
	`user2 --> (Usecase 2)`
	`@enduml`
		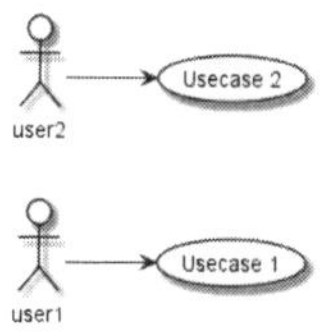

续表

要素	描述	

一个完整的例子	left to right direction
	actor customer
	actor clerk
	rectangle checkout {
	    customer -- (checkout)
	    (checkout) .> (payment) : include
	    (help) .> (checkout) : extends
	    (checkout) -- clerk
	}
		(示意图)

2）类图

类图（Class Diagram）显示了模型的静态结构，特别是模型中存在的类、类的内部结构及它们与其他类的关系等。类图不显示暂时性的信息。类图是面向对象建模的主要组成部分。它既可以用于应用程序的系统分类的一般概念建模，也可以用于详细建模，将模型转换成编程代码，还可以用于数据建模。使用 PlantUML 创建类图的说明如表 1-2 所示。

表 1-2 使用 PlantUML 创建类图的说明

要素	描述		
类的关系	类之间的关系通过下面的符号定义		
	Type	Symbol	Drawing
	Extension（扩展）	<|--	◁—
	Composition（组合）	*--	◆—
	Aggregation（聚合）	o--	◇—
	Implements（实现）	<|..	◁‥
	Association（关联）	-->	←
	Dependency（依赖）	..>	← ‧
	@startuml Class01 <|-- Class02 Class03 <|.. Class04 Class05 --> Class06 newpage Class07 *-- Class08 Class09 o-- Class10 Class11 ..> Class12 @enduml		(示意图)

续表

要素	描述	
关系上的标识	在关系之间使用标签来说明时，使用:连接标签文字 对于元素的说明，用户可以在关系符号的每一边添加说明，并使用 "" 括起来	
	@startuml Class01 "1" *-- "many" Class02 : contains Class03 o-- Class04 : aggregation Class05 --> "1" Class06 @enduml	
添加属性和方法	为了声明字段（对象属性）或方法，可以使用类名连接字段名或方法名，也可以使用{}方式。系统会通过检查是否有括号来判断是方法还是字段	

	@startuml
	class Car
	Car : String type
	Car : void move()
	class Dummy {
	  String data
	  void methods()
	}
	class Flight {
	  flightNumber : Integer
	  departureTime : Date
	}
	@enduml	

定义可访问性	一旦定义了域或方法，就可以定义相应条目的可访问性	

Character	Icon for field	Icon for method	Visibility
-	□	■	private
#	◇	◆	protected
~	△	▲	package private
+	○	●	public

@startuml
class Dummy {
  -field1
  #field2
  ~method1()
  +method2()
}
@enduml

续表

要素	描　　述	
定义可访问性	可以采用以下命令停止使用这些特性 skinparam classAttributeIconSize 0	

	`@startuml`
	`skinparam classAttributeIconSize 0`
	`class Dummy {`
	`　-field1`
	`　#field2`
	`　~method1()`
	`　+method2()`
	`}`
	`@enduml`
		(Dummy 类图：-field1, #field2, ~method1(), +method2())

抽象与静态	使用修饰符{static}或{abstract}可以定义静态或抽象的方法或属性。这些修饰符可以写在行的开始或结束位置	

	`@startuml`
	`class Dummy {`
	`　{static} String id`
	`　{abstract} void methods()`
	`}`
	`@enduml`
		(Dummy 类图：String id, void methods())

抽象类和接口	使用 abstract 或 abstract class 关键字定义抽象类。图形中呈现为斜体。使用 interface、annotation 和 enum 关键字定义接口、标注和枚举	

	`@startuml`
	`abstract class AbstractList`
	`abstract AbstractCollection`
	`interface List`
	`interface Collection`
	`List <|-- AbstractList`
	`Collection <|-- AbstractCollection`
	`Collection <|- List`
	`AbstractCollection <|- AbstractList`
	`AbstractList <|-- ArrayList`
	`class ArrayList {`
	`　Object[] elementData`
	`　size()`
	`}`
	`enum TimeUnit {`
	`　DAYS`
	`　HOURS`
	`　MINUTES`
	`}`
	`annotation SuppressWarnings`
	`@enduml`
		(类图：Collection、List、TimeUnit(DAYS/HOURS/MINUTES)、SuppressWarnings、AbstractCollection、AbstractList、ArrayList(Object[] elementData, size()))

续表

要素	描述	
包	可以通过关键字 package 声明包，同时可选择声明对应的背景色（通过使用 HTML 色彩代码或名称）。注意：包可以被定义为嵌套的	

	@startuml
	package "Classic Collections" #DDDDDD {
	　　Object <|-- ArrayList
	}
	package net.sourceforge.plantuml {
	　　Object <|-- Demo1
	　　Demo1 *- Demo2
	}
	@enduml
		(类图)

1.3.2 数据库建模工具 Workbench

Workbench 是一款专门为 MySQL 设计的集成化桌面软件，也是下一代可视化数据库的设计和管理工具。该软件支持 Windows 和 Linux 系统，可以通过网址 https://dev.mysql.com/downloads/workbench/ 下载。Workbench 为数据库管理员和开发人员提供了一套可视化的数据库操作环境，主要功能有数据库设计与模型建立、SQL 开发（取代 MySQL Query Browser）、数据库管理（取代 MySQL Administrator）。

Workbench 包括开源社区版本和商业版本。
- Workbench Community Edition，也叫 MySQL Workbench OSS，是在 GPL 证书下发布的开源社区版本。
- Workbench Standard Edition，也叫 MySQL Workbench SE，是按年收费的商业版本。

截至本书编写完成时，开源社区版本的最新版本为 MySQL Workbench 8.0.19，其下载界面如图 1-27 所示。

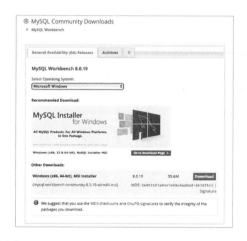

图 1-27　MySQL Workbench 8.0.19 的下载界面

在下载完成后可得到 mysql-workbench-community-8.0.19-winx64.msi 安装文件，双击该文件即可安装。在安装完成后，Workbench 的运行起始界面如图 1-28 所示。

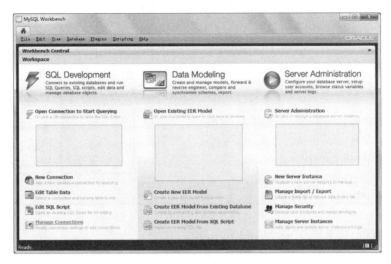

图 1-28　Workbench 的运行起始界面

1.3.3　IDEA 数据库管理工具

IDEA 提供了数据库管理工具，我们可以利用这个工具建立数据库和数据表，操作数据，以及执行 SQL 语句。具体操作方法如下所述。

（1）在第一次使用时，可以选择【View】→【Tool Windows】→【Database】命令，打开数据库管理工具，如图 1-29 所示。然后会在右边栏出现【Database】按钮，以后就可以通过单击右边栏的【Database】按钮来快速打开数据库管理工具，如图 1-30 所示。

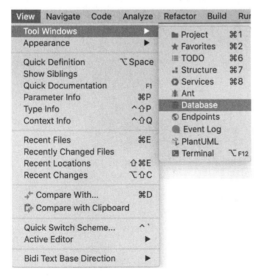

图 1-29　选择命令打开数据库管理工具

图 1-30　快速打开数据库管理工具

（2）单击数据库管理工具左上角的【+】按钮，然后选择【Data Source】→【MySQL】命令，建立 MySQL 数据源，如图 1-31 所示。

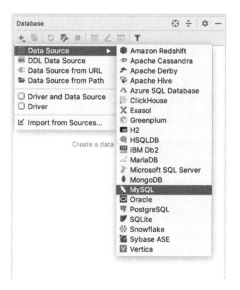

图 1-31　建立 MySQL 数据源

（3）在如图 1-32 所示的界面中进行连接配置，在【Name】文本框中输入一个名称，用来标识这个连接配置，并在下面的几个文本框中分别输入主机名、端口号、用户名和密码，以及数据库名称（如果数据库还未创建，则可以不输入数据库名称），然后单击【OK】按钮。

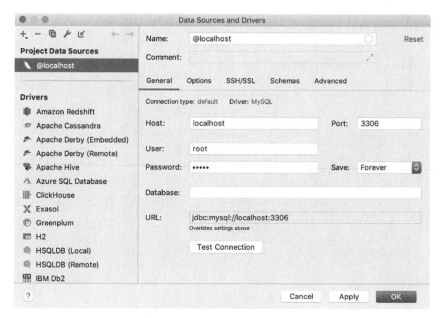

图 1-32　进行连接配置

（4）在如图 1-33 所示的界面左侧输入区中输入创建数据库的 SQL 脚本。

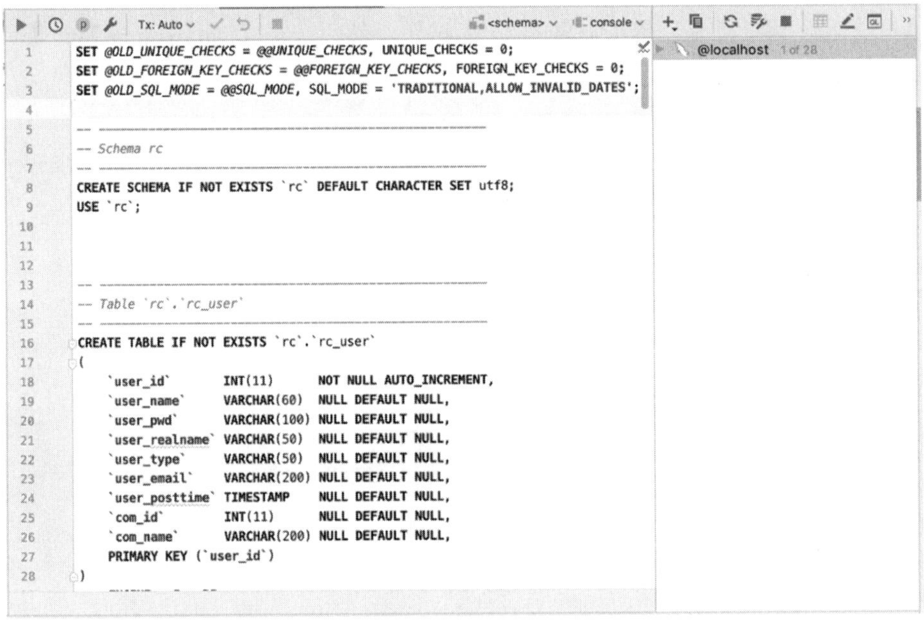

图 1-33 输入 SQL 脚本

（5）选择一段代码，然后单击 ▶ 按钮，在弹出的快捷菜单中选择第二个命令（执行所有 SQL 脚本），如图 1-34 所示。

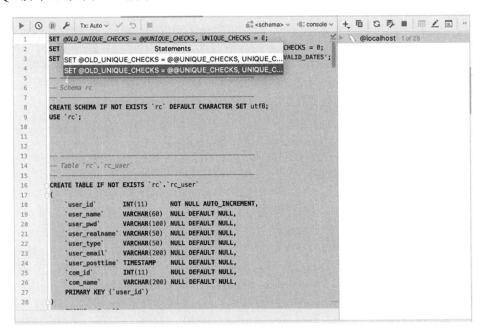

图 1-34 执行 SQL 脚本

（6）在 SQL 脚本执行成功后，即可创建数据库，已建立的数据库如图 1-35 所示。

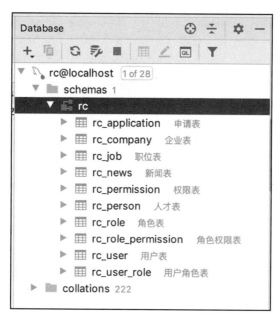

图 1-35 已建立的数据库

1.3.4 基本编码规范与常用技术

编码规范对于软件本身和软件开发人员而言尤为重要。良好的编码规范可以改善软件的可读性,提高团队开发的合作效率,减少软件的维护成本,而且长期的规范性编码可以让开发人员养成良好的编码习惯,甚至锻炼出严谨的思维。

1. 命名规范

(1) 项目名称:项目名称一般采用缩写形式,所有字母均小写。

(2) 包的分类名:包的分类名由小写字母构成,并使用圆点(.)来划分层次。一般格式如下:

[com|org].项目名称.包的分类名

第一部分是可选的,使用 com 表示公司,使用 org 表示组织。第二部分表示项目名称。第三部分表示包的分类名,例如,在按分层架构设计 Web 系统时,控制层、业务层接口、业务实现层、数据访问层接口、数据访问实现层、实体类(模型类)等一般使用 controller、service、service.impl、dao、dao.impl、entity 作为包的分类名。

(3) 文件夹:在视图层划分文件夹可以方便文件的管理。文件夹采用小写英文单词表示。视图层一般按领域模型类来划分文件夹,例如,网上人才中心系统涉及用户、公司、职位、申请、个人等,因此可以建立 user、company、job、application、person 等文件夹,分别用来存放相关的视图文件,还可以建立 images、css、js、common 和 upload 等文件夹,分别用来存放界面图像、样式表、JavaScript 脚本文件、公共网页和上传的文件等。

(4) 文件名:类的文件名与类名相同,扩展名为 java。网页的文件名可根据具体功能来命名,若使用 JSP,则将添加(注册)页面文件、修改页面文件、浏览页面文件、管理页面文件、首页文件分别命名为 add.jsp、edit.jsp、browse.jsp、manage.jsp、index.jsp 等。

（5）类名与接口名：类名由一个或几个英文单词组成，采用大驼峰规则，即每个英文单词的第一个字母大写。类名一般使用完整的英文单词，避免使用缩写词（除非该缩写词被广泛使用，如 Dao）。接口与类的命名类似。按分层结构设计系统，可采用以下命名规范。

- 控制类：项目名+领域模型名+Controller。例如，RcUserController 是网上人才中心系统中用户控制类的名称。
- 模型类（实体类）：项目名+领域模型名。例如，RcUser 是网上人才中心系统中用户模型类的名称。
- 业务逻辑接口：项目名+领域模型名+Service。例如，RcUserService 是网上人才中心系统中用户业务逻辑接口的名称。
- 业务逻辑类：项目名+领域模型名+Service+Impl。例如，RcUserServiceImpl 是网上人才中心系统中用户业务逻辑类的名称。
- 数据访问接口：项目名+领域模型名+Dao。例如，RcUserDao 是网上人才中心系统中用户数据访问接口的名称。
- 数据访问类：项目名+领域模型名+Dao+Impl。例如，RcUserDaoImpl 是网上人才中心系统中用户数据访问类的名称。

（6）方法名：方法名的第一个英文单词的首字母小写，其他英文单词的首字母大写，如 getPersonInfo()。此外，还有一些规范如下所述。

- 属性方法遵循 JavaBean 命名规范：getXXX()和 setXXX()。
- 转换对象类型的方法一般被命名为 toType()形式，如 toString()。
- 返回 boolean 类型的方法被命名为 isXXX()形式，如 isTriangle()。
- 数据访问层的方法应尽量体现 SQL 操作。例如，将添加、修改、删除、按 ID 查询、根据条件查一个、根据条件查多个、查所有、按条件分页查询、查所有总数、按条件查总数的方法分别命名为 insert、update、delete、selectById、selectOne、selectSome、selectAll、selectSomeByPage、selectAllCount、selectCount。
- 业务逻辑层的方法应尽量反映业务功能，要求可读性强。例如，对于用户的业务逻辑层而言，将添加、修改、删除、按 ID 查询、查所有、按条件分页查询、查所有总数、按条件查总数的方法分别命名为 addUser、editUser、deleteUser、findById、findAllUsers、findUsersByPage、findAllCount、findCount。
- 控制类的方法反映了客户端请求，应将添加、登录、要修改、修改、删除、管理、浏览、查看等请求的方法分别命名为 add、login、willEdit、edit、delete、manage、browse、show 等。

（7）变量名与常量名：变量名一般采用小驼峰规则，即第一个英文单词的首字母小写，其后的英文单词的首字母大写。常量名由一个或多个被下画线分开的大写英文单词组成，如 PAGE_SIZE。

2. 注释规范

注释是在源程序中具有说明作用的语句，这种语句在编译时将被编译器忽略。注释是程序设计中不可或缺的组成部分，目的是增加程序的可读性。需要说明的是，因为本书的篇幅限制，书中所提供的代码省去或简化了注释。Java 的注释包括以下 3 种。

- 单行注释：以//开头，直到行的末尾。
- 多行注释：以/*开头，直到*/结束，用来注释一行或多行。
- 文档注释：以/**开头，直到*/结束，这是 Java 特有的注释方法，能被转化为 HTML 格式的帮助文档。

在 Java EE 程序设计中，注释一般遵循以下规范。

（1）源文件注释：源文件注释采用/*…*/，在每个源文件的头部，主要用于描述文件名、版权信息等。

（2）类注释：类注释采用/** … */，在类的前面，主要用于描述类的作用、版本、可运行的 JDK 版本、作者、时间等。可以使用的标记包括以下几种。

- @author：描述作者。
- @version：描述版本。
- @since：描述该类可以运行的 JDK 版本。
- @datetime 描述该类建立的时间。

（3）方法注释：方法注释采用/**… */，在方法的前面，主要用于描述方法的功能、参数、返回值、异常等。可以使用的标记包括以下几种。

- @param：描述方法的参数。
- @return：描述方法的返回值。对于无返回值的方法或构造方法而言，@return 可以被省略。
- @throws：描述在什么情况下抛出什么类型的异常。

（4）全局变量注释：如果是 public 类型的变量或常量，则应使用/**…*/注释对其进行重点说明。如果是其他类型的变量，则可以使用//注释进行简单说明，但是需要在它的设置（set 方法）与获取（get 方法）成员方法上加上方法注释。

（5）内部代码注释：方法内部的代码使用/*…*/或//进行注释，主要用于对代码进行一些必要的说明。

3．使用 IDEA 内置的功能生成注释

1）类注释模板

设置类注释模板的过程如下所述。

（1）选择【File】→【Setting】命令，打开配置对话框。

（2）选择【Editor】→【File and Code Templates】→【Class】选项，打开类注释的模板。

（3）在 public class ${NAME} {的上一行插入自定义的模板，如图 1-36 所示。

```
/**
 * 名称：${NAME}
 * 描述：${description}
 * @author: Yang ShuLin
 * @datetime: ${YEAR}-${MONTH}-${DAY} ${HOUR}:${MINUTE}
 * @version 1.0
 */
```

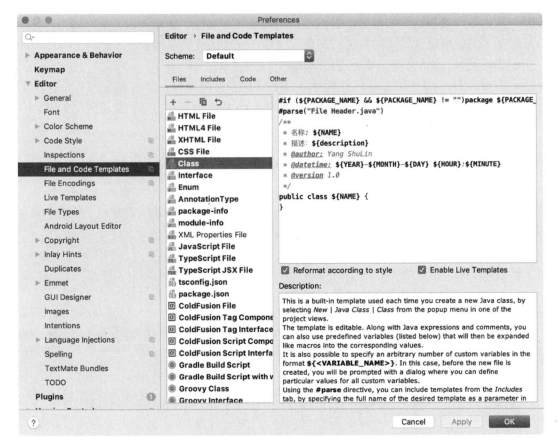

图 1-36　插入自定义的模板

上述 7 个变量说明如下所述。

- ${NAME}：类的名称。
- ${description}：在创建类时，提示输入类的描述。
- ${YEAR}：完整的年份，如 2020。
- ${MONTH}：完整的月份，如 02。
- ${DAY}：完整的日期，如 22。
- ${HOUR}：24 小时制的小时，如 16。
- ${MINUTE}：完整的分钟，如 52。

在设置好后，创建新类时会弹出对话框，按提示输入描述的内容，就会自动生成注释。

2）方法注释模板

设置方法注释模板的过程如下所述。

（1）选择【File】→【Setting】命令，打开配置对话框。

（2）选择【Editor】→【Live Templates】选项，打开实时模板设置界面。

（3）单击界面右侧的绿色加号按钮，在弹出的下拉列表中选择【Template Group】选项，创建一个自定义的模板组，如图 1-37 所示。

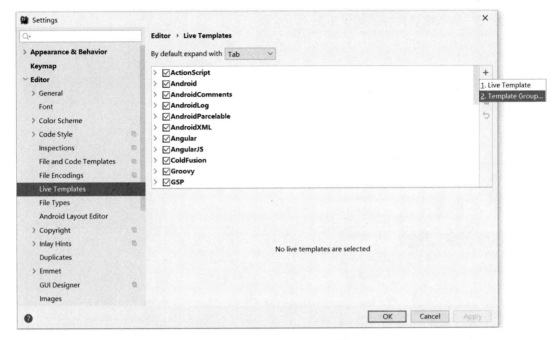

图 1-37　创建一个自定义的模板组

（4）选择刚刚创建的模板组，单击界面右侧的绿色加号按钮，在弹出的下拉列表中选择【Live Template】选项，在该模板组下创建一个自定义的模板，如图 1-38 所示。

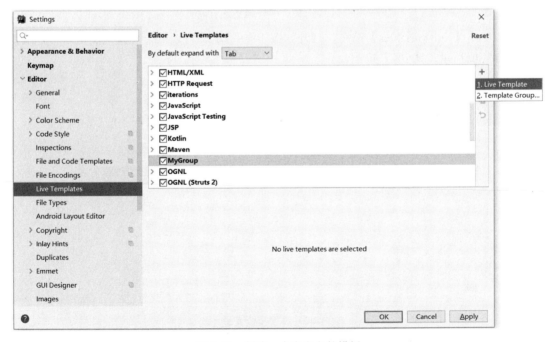

图 1-38　创建一个自定义的模板

（5）选择刚刚创建的模板，在下方的【Abbreviation】文本框中输入【*】，并在【Template

text】列表框中输入模板内容，如图 1-39 所示。

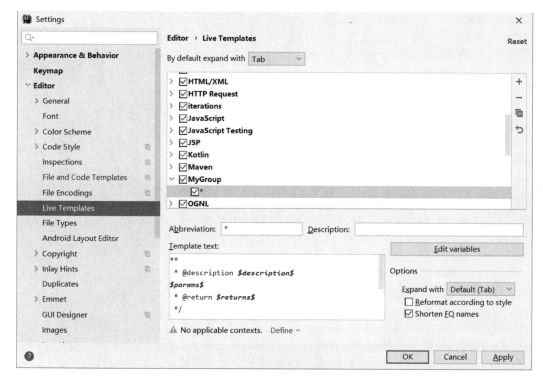

图 1-39　编辑模板

（6）单击【Edit variables】按钮，打开如图 1-40 所示的对话框，对模板变量进行设置，并在设置完成后单击【OK】按钮。

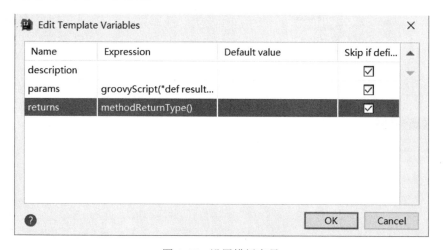

图 1-40　设置模板变量

其中，description 的【Expression】栏保持为空。

在 params 的【Expression】栏中输入：

groovyScript("def result=''; def params=\"${_1}\".replaceAll('[\\\\[|\\\\]|\\\\s]', '').split(',').toList(); for(i = 0; i < params.size(); i++) {if(params[i] == '') return result; result+=' * @param ' + params[i] + ((i < params.size() - 1) ? '\\n' : '')}; return result", methodParameters())

在 returns 的【Expression】栏中选择【methodReturnType()】选项。

（7）返回如图 1-39 所示的界面，单击界面下方【No applicable contexts】旁边的【Define】，在弹出的下拉列表中勾选【Java】复选框，选择可用的环境，如图 1-41 所示。然后单击其他地方关闭下拉列表，并单击【OK】按钮完成模板设置。

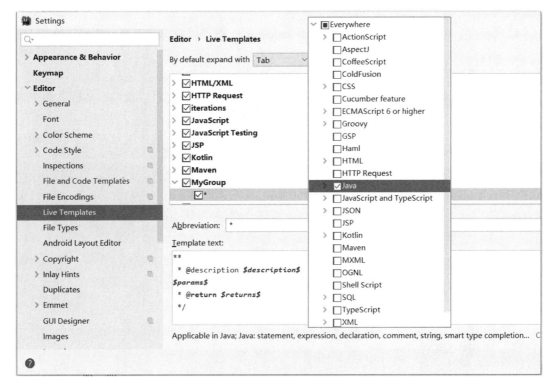

图 1-41　选择可用的环境

若要使用设置好的模板，则只需要在方法前输入【/*】并按回车键即可。

1.3.5　Spring Boot 集成 Swagger2

在团队开发中，一个好的 API 文档不但可以大大降低沟通成本，而且可以帮助新人快速上手业务。传统的做法是由开发人员创建一份 RESTful API 文档来记录所有的接口细节，并在程序员之间互相传递。这种做法存在以下几个问题。

（1）API 接口众多，细节复杂，需要考虑不同的 HTTP 请求类型、HTTP 头部信息、HTTP 请求内容等。若想要高质量地完成这份文档，需要耗费大量的精力，并且难以维护。

（2）随着需求的变更和项目的优化、推进，接口的细节会不断地演变，因此接口描述文档也需要同步修订，但是文档和代码处于两个不同的媒介中，除非有严格的管理机制，否则很容易出现文档、接口不一致的情况。

Swagger2 的出现就是为了从根本上解决上述问题。

1. Swagger2 及其配置

Swagger 是一组围绕 OpenAPI 规范构建的开源工具，可以帮助用户设计、构建、记录和使用 RESTful API。作为一个规范和完整的框架，Swagger 可以用于生成、描述、调用和可视化 RESTful 风格的 Web 服务。接口文档可以在线自动生成，并随接口变动实时更新，从而节省维护成本。同时接口文档支持在线接口测试，不依赖第三方工具。Swagger2 是 Swagger 的当前版本，具有以下特点。

- 代码变，文档变。只需要少量的标注，Swagger 就可以根据代码自动生成 API 文档，很好地保证了文档的时效性。
- 跨语言性，支持 40 多种语言。
- Swagger UI 呈现出来的是一份可交互式的 API 文档，我们可以直接在文档页面中尝试调用 API，省去了准备复杂的调用参数的过程。
- 可以将文档规范导入相关的工具（如 Postman、SoapUI），这些工具将会创建自动化测试。

在项目中使用 Swagger2，可进行以下配置。

1）引入依赖

```
<dependency>
    <groupId>io.springfox</groupId>
    <artifactId>springfox-swagger2</artifactId>
    <version>2.9.2</version>
</dependency>
<dependency>
    <groupId>io.springfox</groupId>
    <artifactId>springfox-swagger-ui</artifactId>
    <version>2.9.2</version>
</dependency>
```

2）创建配置类

```
@Configuration
@EnableSwagger2
@EnableWebMvc
public class SwaggerConfig implements WebMvcConfigurer {
    @Bean
    public Docket createRestApi() {
        return new Docket(DocumentationType.SWAGGER_2)
                .apiInfo(apiInfo())
                .select()
                //为当前包下的 Controller 生成 API 文档
                //.apis(RequestHandlerSelectors.basePackage("com.troila"))
                //为有@Api 标注的 Controller 生成 API 文档
                //.apis(RequestHandlerSelectors.withClassAnnotation(Api.class))
                //为有@ApiOperation 标注的方法生成 API 文档
                //.apis(RequestHandlerSelectors.withMethodAnnotation(ApiOperation.class))
```

```
                //为任何接口生成 API 文档
                .apis(RequestHandlerSelectors.any())
                //.apis(RequestHandlerSelectors.basePackage("com.cjs.res.controller"))
                .paths(PathSelectors.any())
                .build();
    }
    private ApiInfo apiInfo() {
        return new ApiInfoBuilder()
                .title("example 系统接口文档")
                .description(initContextInfo())
                //服务条款网址
                //.termsOfServiceUrl("")
                .contact(new Contact("XXXX", "http://www.baig", "fff@bbb.com"))
                .version("1.0")
                .build();
    }

    private String initContextInfo() {
        StringBuffer sb = new StringBuffer();
        sb.append("RESTful API 设计在细节上有很多自己独特的需要注意的技巧, 并且对开发人员的架构设计能力比传统 API 有着更高的要求。")
            .append("<br/>")
            .append("本文通过翔实的叙述和一系列的范例, 从整体结构到局部细节, 分析和解读了为了提高易用性和高效性, RESTful API 设计应该注意哪些问题, 以及如何解决这些问题。");
        return sb.toString();
    }
    @Override
    public void addResourceHandlers(ResourceHandlerRegistry registry) {
        registry.addResourceHandler("swagger-ui.html")
                .addResourceLocations("classpath:/META-INF/resources/");
        registry.addResourceHandler("/webjars/**")
                .addResourceLocations("classpath:/META-INF/resources/webjars/");
    }
}
```

上述类使用@Configuration 标注，表明它是一个配置类；@EnableSwagger2 表示开启 Swagger2。

在通过 createRestApi()方法创建 Docket 的 Bean 之后，apiInfo()方法用来创建该 API 的基本信息（这些基本信息会展现在文档页面中）；select()方法用来返回一个 ApiSelectorBuilder 实例，控制哪些接口会暴露给 Swagger；apis()方法用来指定扫描的包以生成文档展现。本例指定了包路径，Swagger 会扫描该包下所有 Controller 定义的 API，并产生文档内容。

如果是 Spring Boot 与 JSP 结构的程序，还需要继承 WebMvcConfigurer，并通过覆盖 addResourceHandlers 设置路径。

2. 使用 Swagger2

在项目中使用 Swagger2 主要通过以下标注。

- @Api：修饰整个类，描述 Controller 的作用。

- @ApiOperation：描述一个方法。
- @ApiImplicitParam：描述一个参数，可以配置参数的中文含义，也可以给参数设置默认值，从而在接口测试时可以避免手动输入。
- @ApiImplicitParams：如果有多个参数，则需要使用多个@ApiImplicitParam 标注来描述，多个@ApiImplicitParam 标注需要放在一个@ApiImplicitParams 标注中。
- @ApiModel：描述模型类。
- @ApiModelProperty：描述模型类的属性。
- @ApiIgnore：用于方法或类，表示这个方法或类被忽略。
- @ApiError：描述发生错误时返回的信息。
- @ApiResponse：HTTP 响应其中一个描述。
- @ApiResponses：HTTP 响应整体描述。

1）类描述

@Api 用于描述类，说明该类的作用。可以标记一个 Controller 类作为 Swagger 文档资源，使用方式如下：

@Api(value = "/user", description = "请求用户控制类")

2）方法描述

@ApiOperation 用于描述一个方法，基本格式如下：

@ApiOperation(value = "接口说明", httpMethod = "接口请求方式", response = "接口返回参数类型", notes = "接口发布说明")

例如：

@ApiOperation(value = "根据用户名获取用户的信息",notes = "查询数据库中的记录",httpMethod = "POST",response = String.class)

3）方法各参数描述

单个参数用@ApiParam 描述，基本格式如下：

@ApiParam(required = "是否为必需参数", name = "参数名称", value = "参数具体描述",dateType ="变量类型",paramType="请求方式")

例如：

@ApiParam(name = "userId",value = "用户 ID",required = true,dataType = "Integer",paramType = "query")

多个参数用@ApiImplicitParams 描述，并通过@ApiImplicitParam 对每个参数进行描述，基本格式如下：

@ApiImplicitParams({
　　@ApiImplicitParam(…),
　　@ApiImplicitParam(…),
　　…
})

@ApiImplicitParam 的基本格式如下：

@ApiImplicitParam(required = "是否为必需参数", name = "参数名称", value = "参数具体描述",dateType="变量类型",paramType="请求方式")

例如：

@ApiImplicitParams({

```
        @ApiImplicitParam(name = "nickName",value = "用户的昵称",required = true, dataType = 
"String", paramType = "query"),
        @ApiImplicitParam(name = "id",value = "用户的 ID" ,required = true,dataType = "Integer", 
paramType = "query")
})
public String getUserInfoByNickName(String nickName, Integer id) {
    return "1234";
}
```

响应结果用@ApiResponse 描述，基本格式如下：
@ApiResponse(code="http 的状态码"，message="描述")
多个响应用@ApiResponses 描述。例如：
```
@ApiResponses(value = {
        @ApiResponse(code = 200, message = "Successful — 请求已完成"),
        @ApiResponse(code = 400, message = "请求中有语法问题，或不能满足请求"),
        @ApiResponse(code = 401, message = "未授权客户机访问数据"),
        @ApiResponse(code = 404, message = "服务器找不到给定的资源；文档不存在"),
        @ApiResponse(code = 500, message = "服务器不能完成请求")}
)
```

1.4 Java EE 项目的分层架构模式

在传统的系统设计中，会将数据库的访问、业务逻辑及可视元素等代码混杂在一起，这样虽然直观，但是代码的可读性差、耦合度高，会为日后的维护和重构带来不便。为了解决这个问题，有人提出了分层架构思想，即将各个功能分开，放在独立的层中，使各层通过协作来实现整体功能。使用分层架构设计容易实现如下目的：分散关注、松散耦合、逻辑复用、标准定义。

1.4.1 分层架构模式概述

分层（Layer）模式是非常常见的一种架构模式，甚至可以说，分层模式是很多架构模式的基础。分层描述的是这样一种架构设计过程：从最低级别的抽象开始，称为第 1 层，这是系统的基础；通过将第 J 层放置在第 $J-1$ 层的上面来逐步向上完成抽象阶梯，直到到达功能的最高级别，称为第 N 层。

因此分层模式可以定义为：将解决方案的组件分隔到不同的层中，同时每一层中的组件应保持内聚性，并且应大致在同一抽象级别，每一层都应与它下面的各层保持松散耦合。

分层模式的关键在于确定依赖，即通过分层可以限制子系统间的依赖关系，使系统以更松散的方式耦合，从而更易于维护。

- 伸缩性：伸缩性指的是应用程序是否能支持更多的用户。应用的层数越少，则可以增加资源（如 CPU 和内存）的地方就越少；层数越多，则可以将每一层分布在不同的机器上。

- 可维护性：可维护性指的是当发生需求变化时，只需修改软件的某一部分代码，不会影响其他部分的代码。
- 可扩展性：可扩展性指的是在现有系统中增加新功能的难易程度。层数越多，则可以在每一层中提供扩展点，不会打破应用的整体框架。
- 可重用性：可重用性指的是程序代码没有冗余，同一个程序能满足多种需求。例如，业务逻辑层可以由多种表示层共享。
- 可管理性：可管理性指的是管理系统的难易程度。在将应用程序分为多层后，可以将工作分解给不同的开发小组，从而便于管理。应用越复杂，规模越大，则需要的层就越多。

在分层架构的设计中，需要遵循以下原则。

- 单向逐层调用原则：现在约定将 N 层架构的各层依次编号为 1、2、…、K、…、$N-1$、N，其中层的编号越大，则越处在上层。同时要求第 K（$1<K\leq N$）层只可依赖第 $K-1$ 层，而不可依赖其他层。那么，如果第 P 层依赖第 Q 层，则 P 一定大于 Q。
- 面向接口编程原则：接口是一组规则的集合，它规定了实现本接口的类或接口必须拥有的一组规则，体现了自然界"如果你是……则必须能……"的理念。现在仍约定将 N 层架构的各层依次编号为 1、2、…、K、…、$N-1$、N，其中层的编号越大，则越处在上层。同时要求第 K 层不应该依赖具体的第 $K-1$ 层，而应该依赖一个第 $K-1$ 层的接口，即在第 K 层中不应该有第 $K-1$ 层中的某个具体类。
- 封装变化原则：找出应用中可能需要变化的代码，并将它们独立出来，避免它们和那些不需要变化的代码混杂在一起。
- 开闭原则：对扩展开放，对修改关闭。具体到 N 层架构中，可以描述为：当第 $K-1$ 层有一个新的具体实现时，它应该可以在不修改第 K 层的情况下，与第 K 层无缝连接，顺利交互。
- 单一职责原则：任何一个类都应该有单一的职责，属于单独的一层，不能同时担负两种职责或属于多个层。
- 接口平行原则：某一实体对应的接口组应该是平级且平行的，不应该跨越多个实体或多个级别。

1.4.2　Java Web 应用中的三层结构

Java EE 项目以 Java Web 应用为主。在 Java Web 应用系统开发中，比较流行三层结构（不包括后台数据库），即将系统分为表示层、业务逻辑层和数据访问层。

- 表示层：表示层位于最外层（最上层），离用户最近。该层用于显示数据和接收用户输入的数据，为用户提供一种交互式操作的界面。该层会对流入的数据的正确性和有效性负责，对呈现样式负责，对呈现的错误信息负责。

第 1 章　Java EE 与企业级应用开发

- 业务逻辑层：业务逻辑层处于数据访问层与表示层中间，在数据交换中具有承上启下的作用。由于层是一种弱耦合结构，层与层之间的依赖是向下的，下层对于上层而言是"无知"的，改变上层的设计对于其调用的下层而言没有任何影响。如果在分层设计时遵循了面向接口设计的思想，则这种向下的依赖也应该是一种弱依赖关系。因此在不改变接口定义的前提下，理想的分层式架构应该是一个支持可抽取、可替换的"抽屉"式架构。而且业务逻辑层的设计对一个支持可扩展的架构尤为关键，因为它扮演了两个不同的角色：对于数据访问层而言，业务逻辑层是调用者；对于表示层而言，业务逻辑层却是被调用者。依赖与被依赖的关系都处于业务逻辑层。业务逻辑层不仅负责系统领域业务的处理，还负责逻辑性数据的生成、处理及转换。
- 数据访问层：数据访问层有时也被称为持久化层，主要负责数据库的访问，可以访问数据库系统、二进制文件、文本文档或 XML 文档。简单来说，数据访问层可以实现对数据表的 Select、Insert、Update、Delete 等操作。如果要加入 ORM（Object Relation Mapping）的元素，数据访问层就会包括对象和数据表之间的 Mapping（映射），以及对象实体的持久化。数据访问层不负责数据的正确性和可用性，不了解数据的用途，不负责任何业务逻辑。

1.4.3　结合 MVC 模式的分层结构

MVC 思想表示将一个应用分成 Model（模型）、View（视图）、Control（控制）3 个模块。这 3 个模块以最少的耦合协同工作，从而提高应用的可扩展性和维护性。其中，模型模块用于实现系统中的业务逻辑，通常用 JavaBean 或 EJB 来实现。视图模块用于与用户的交互，通常用 JSP 或 JSF 来实现。控制模块是模型模块与视图模块之间沟通的桥梁，它可以分配用户的请求并选择恰当的视图来显示，同时它可以解释用户的输入并将它们映射为模型模块可执行的操作。按这种模式设计程序，多个视图模块可以对应一个模型模块，模型模块返回的数据可以与显示逻辑分离，从而使得程序结构清晰、易于维护。

在前文所述的三层结构中，业务逻辑层具有关键的作用，它隔离了表示层和数据访问层，使系统易于维护和扩展。但是，这样的分层结构并没有解决表示层的问题。在表示层中，处理用户的请求、调用业务功能、显示界面等功能代码还混杂在一起，并且在 JSP 中还经常嵌入 Java 代码，使这部分代码难以重用，导致程序的结构不清晰。为此，结合 MVC 模式，可以将三层结构中的表示层进一步划分为视图层和控制层，使得页面与控制逻辑分离，程序结构清晰，便于重用和维护。这种结合 MVC 模式的分层结构的具体对应关系如图 1-42 所示。

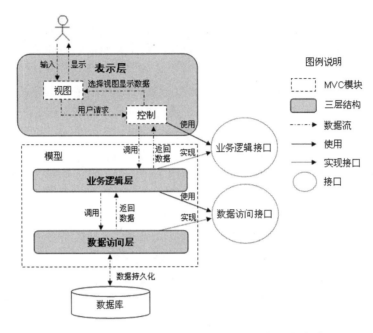

图 1-42　结合 MVC 模式的分层结构的具体对应关系

1.4.4　网上人才中心系统分析与设计

在网络经济环境下，网络招聘以其效率高、成本低、覆盖面广等优势显示出了巨大的发展潜力。网上人才中心系统的设计目的就是为企业和人才搭建一个桥梁，并借助网络，实现企业和人才的交互选择。

1. 需求描述

网上人才中心系统的主要功能就是让个人用户（人才）通过网络快速找到自己满意的工作，让企业通过网络发布和管理招聘信息，对应聘者进行回复。网上人才中心系统主要提供以下功能。

- 个人用户和企业用户均可在系统中注册，并在注册后可登录。对不同的用户显示不同的状态。
- 在企业用户登录后，可以发布招聘信息、管理应聘信息、回复应聘者，以及浏览、查询和查看人才信息。
- 在个人用户登录后，可以修改个人信息、查询和查看企业信息、查询职位、申请职位（即应聘）、管理个人申请。
- 在管理员登录后，可以管理企业、管理用户、管理权限、管理角色、管理新闻，并修改管理员密码。
- 系统能够验证用户的权限，使得不同的用户具有不同的权限。

2. 用例分析

用例图用于显示外部参与者与系统的交互，能够更直观地描述系统的功能。从角色来看，网上人才中心系统的用户分为个人用户、企业用户和管理员。如图1-43、图1-44和图1-45所示为网上人才中心系统的3个用例图。

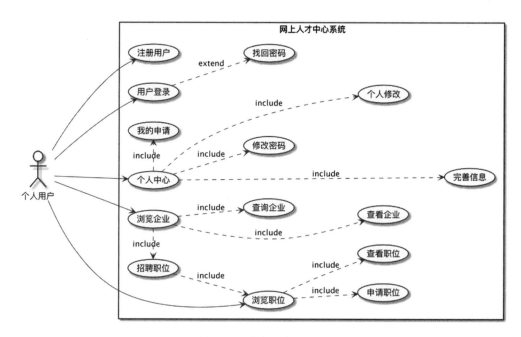

图1-43 个人用户用例图

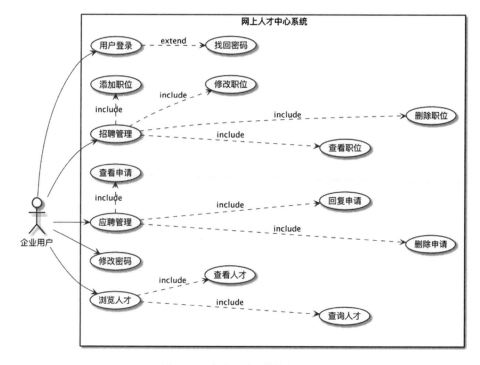

图1-44 企业用户用例图

Java EE 企业级应用开发技术研究

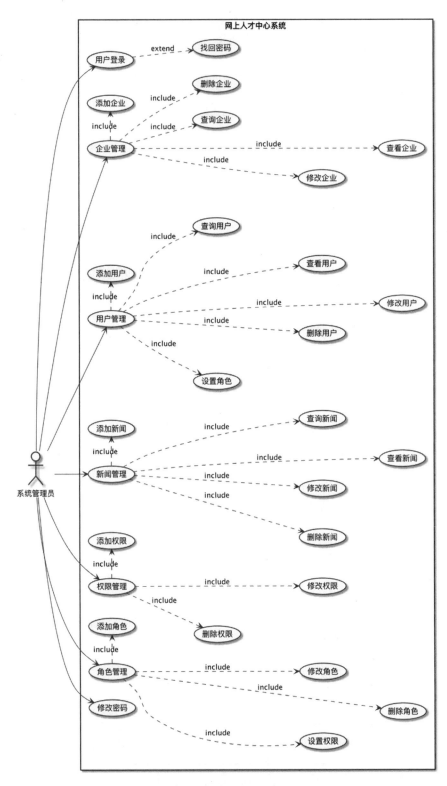

图 1-45 管理员用例图

3. 功能描述

网上人才中心系统的功能划分如表 1-3 所示。

表 1-3 网上人才中心系统的功能划分

模块名	子功能	描述
系统首页	登录/注册	允许个人用户注册，允许个人用户、企业用户或管理员登录
	站内新闻	能够浏览、查看站内新闻
	职位信息	登录的个人用户能够浏览、查询、查看招聘的职位，并申请职位（应聘）
	企业信息	登录的个人用户能够浏览、查询、查看企业信息，并查看企业招聘的职位
	人才信息	登录的企业用户能够浏览、查询、查看人才信息
个人中心	个人中心	进入个人中心
	个人修改	能够对个人用户注册的信息进行修改
	完善信息	能够设置个人用户作为人才的信息
	我的申请	能够浏览、查看、取消自己应聘的职位，查看企业的回复
	修改密码	个人用户可修改密码
企业管理	信息修改	能够对企业用户注册的信息进行修改
	招聘管理	能够浏览、查询、添加、修改、查看、删除招聘的职位信息
	应聘管理	能够浏览、回复、删除应聘信息，查看申请者
	修改密码	企业用户可修改密码
系统管理	企业管理	能够浏览、查询、删除、查看企业用户信息
	用户管理	能够浏览、查询、添加、修改、删除、查看用户信息，设置用户权限
	权限管理	能够浏览、添加、修改、删除权限
	角色管理	能够浏览、添加、修改、删除角色，设置角色权限
	新闻管理	能够浏览、添加、修改、删除新闻
	修改密码	管理员可修改密码

4. 其他需求

根据用户对本系统的需求，系统应当在响应时间、可靠性、安全性等方面具有较高的性能。

1）界面需求

系统的界面需求如下所述。

- 页面内容：主题突出，栏目、菜单的设置和布局合理，传递的信息准确、及时；内容丰富，文字准确，语句通顺；站点定义、术语和行文格式统一、规范、明确。
- 导航结构：页面具有明确的导航指示，并且便于理解，方便用户使用。
- 技术环境：页面大小适当，可以使用各种常用浏览器以不同分辨率浏览；无错误链接和空链接；采用 CSS 处理，控制字体大小和版面布局。
- 艺术风格：界面、版面形象清新悦目、布局合理；字号大小适宜、字体选择合理，前后一致，美观大方；动与静搭配恰当，动静效果好；色彩和谐自然，与主题内容协调。

2）响应时间需求

用户在系统中进行任何操作时,系统都应当及时地进行反应,反应的时间应控制在 5 秒以内。系统应当能够监测出各种非正常情况,如与设备的通信中断,无法连接数据库服务器等,避免出现让用户长时间等待甚至无响应的情况。

3）可靠性需求

系统应当保证 2000 人可以同时在客户端登录,并正常运行,正确提示相关内容。

4）开放性需求

系统应当具有较大的灵活性,以适应将来功能扩展的需求。

5）可扩展性需求

系统应当具有可扩展性,以适应将来功能扩展的需求。

6）系统安全性需求

系统应当具有严格的权限管理功能,其各功能模块需要用户具有相应的权限才能进入。系统应当能够防止各类误操作可能造成的数据丢失、损坏等情况,防止用户非法获取网页及内容。

5. 系统设计

1）系统功能结构

网上人才中心系统的功能结构如图 1-46 所示。

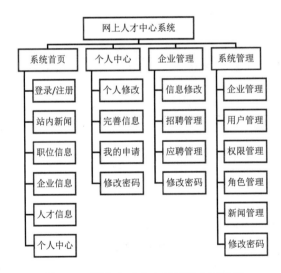

图 1-46 网上人才中心系统的功能结构

2）数据库设计

根据企业信息展示系统的要求,主要涉及用户、角色、权限、个人、企业、招聘、应聘、新闻等数据,因此建立 10 个数据表用来存储对应的数据。

将系统数据库命名为 rc,并将 10 个数据表分别命名为 rc_user（用户）、rc_person（个人）、rc_role（角色）、rc_permission（权限）、rc_user_role（用户角色）、rc_role_permission（角色权限）、rc_company（企业）、rc_job（工作）、rc_application（申请）、rc_news（新闻）。

各个数据表及其关系如图 1-47 所示。

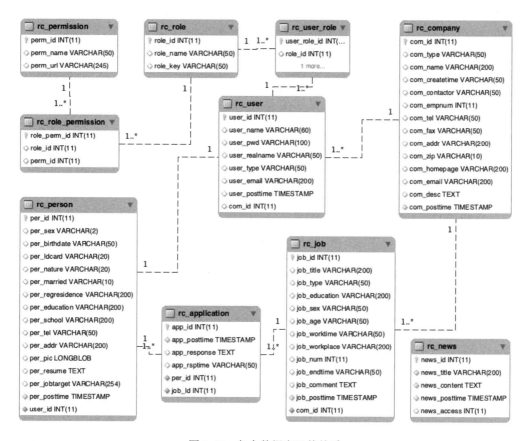

图 1-47　各个数据表及其关系

3）实体类设计

网上人才中心系统主要包含如下实体类（也称为数据模型类）：RcUser（用户类）、RcRole（角色类）、RcPermission（权限类）、RcPerson（个人）、RcCompany（企业）、RcJob（工作，即职位）、RcApplication（申请，即应聘）、RcNews（新闻）。这些类均被存放在 com.rc.entity 包下。UML 类图如图 1-48 所示，该类图隐藏了属性方法。

在建立这些类时，需要注意以下几点。

（1）每个实体类都要有一个无参数的构造函数。

（2）每个实体类要实现 Serializable 接口。

（3）RcJob 类和 RcCompany 类是多对一单向关联的关系。

（4）RcApplication 类和 RcJob 类及 RcPerson 类均是多对一单向关联的关系。

（5）RcPerson 类和 RcUser 类，以及 RcCompany 类和 RcUser 类均是多对一单向关联的关系。

4）业务逻辑层接口设计

业务逻辑层接口体现了业务功能的需要，反映了系统所具有的业务功能。业务逻辑层接口主要包括 RcUserService（用户业务逻辑接口）、RcRoleService（角色业务逻辑接口）、RcPermissionService（权限业务逻辑接口）、RcPersonService（个人业务逻辑接口）、RcCompanyService（企业业务逻辑接口）、RcJobService（职位业务逻辑接口）、RcApplicationService（申请业务逻辑接口）、RcNewsService（新闻业务逻辑接口）。这些接口均被存放在 com.rc.service 包下。业务逻辑层接口图如图 1-49 所示。

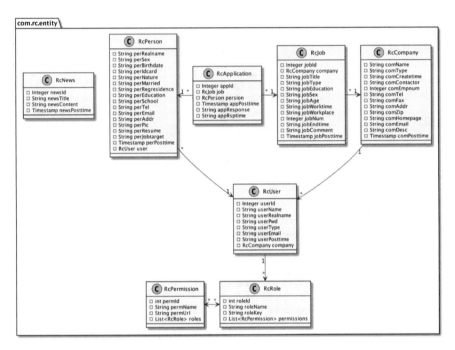

图 1-48　UML 类图

图 1-49　业务逻辑层接口图

5）数据访问层接口设计

数据访问层用于实现对数据库的访问，其接口定义了数据访问的方法。数据访问层接口主要包括 RcUserDao（用户数据访问接口）、RcRoleDao（角色数据访问接口）、

RcPermissionDao（权限数据访问接口）、RcPersonDao（个人数据访问接口）、RcCompanyDao（企业数据访问接口）、RcJobDao（职位数据访问接口）、RcApplicationDao（申请数据访问接口）、RcNewsDao（新闻数据访问接口）。这些接口均被存放在 com.rc.dao 包下。数据访问层接口图如图1-50所示。

图1-50 数据访问层接口图

6）总体界面设计

为了使网页具有代码简练、容易重构、访问速度快、搜索引擎友好、浏览器兼容性好等特点，界面的结构布局主要使用 DIV 标签并与 CSS 结合来实现。

（1）绘制界面效果图。

使用 Photoshop 等绘图工具绘制界面效果图，如图1-51所示。

图1-51 网上人才中心系统界面效果图

（2）系统界面结构分析。

系统界面有 4 个主要区域，即题头区（header）、菜单（menu）、主体区（pagebody）及页脚区（footer），其中主体区包含左区（left）和右区（right）。系统界面的结构布局如图 1-52 所示。

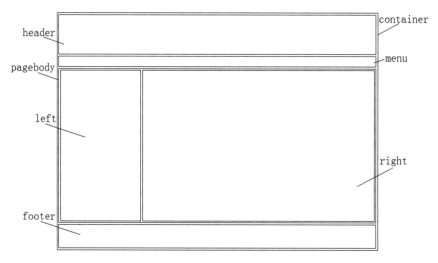

图 1-52　系统界面的结构布局

（3）确定 DIV 标签的层次及命名。

DIV 标签的层次及命名如图 1-53 所示。

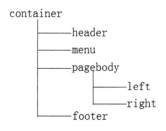

图 1-53　DIV 标签的层次及命名

（4）设计系统首页。

使用 HTML 进行系统首页设计，代码如下：

```
<!DOCTYPE html>
<html>
<head>
    <meta http-equiv="Content-Type" content="text/html; charset=UTF-8">
    <title>网上人才中心系统</title>
    <link rel="stylesheet" href="/rc/static/css/all.css" type="text/css"/>
</head>
</head>
<body>
<div id="container">
```

```html
        <div id="header"></div>
        <div id="menu">
            <ul>
                <li><a href="/rc ">【系统首页】</a></li>
                <li><a href="/rc/news/list" target="content">【站内新闻】</a></li>
                <li><a href="/rc/job/list" target="content">【职位信息】</a></li>
                <li><a href="/rc/company/list" target="content">【企业信息】</a></li>
                <li><a href="/rc/person/list" target="content">【人才信息】</a></li>
                <li><a href="/rc/person/center" target="content">【个人中心】</a></li>
            </ul>
        </div>
        <div id="pagebody">
            <div id="left" style="text-align: center">
                左区
            </div>
            <div id="right">
                右区
            </div>
        </div>
        <div id="footer">
            <hr/>
            Copyright &copy; 2020-2030 YSL.All Rights Reserved.
        </div>
    </div>
</body>
</html>
```

（5）建立样式表文件 all.css。

在 css 文件夹下建立 all.css 文件，并编写代码如下：

```css
@charset "utf-8";
body {
    margin:0; text-align:center; font-size:14px; font-family:宋体,arial;
}
input,select,textarea,table{
    font-size:14px;
}
a{
    font-size:14px; color:#000080; text-decoration:none;
}
a:hover{
    color:#0000ff;
}
p{
    clear:both; margin:5px 0 0 0; padding:0 0 0 10px;
}
#container{
    width:1024px; margin:0px auto; background-color:#66FFFF;
}
#header{
    height:146px; background:url(../images/main.gif) 0 0 repeat-x;
```

```css
        border-bottom:1px solid #999999;
}
#menu{
        background-color: #00FFFF; width:100%;
}
#menu ul{
        margin:0px auto; padding:0px; list-style-type:none; height:30px; width: 720px;
        line-height:30px; text-align:center;
}
#menu ul li{
        width:120px; float:left; text-align:center;
}
#pagebody{
        clear:both; width:100%; min-height:300px; border-top:1px solid #999999;
        text-align:center; font-size:14px;
}
#pagebody form p{
        clear:both; margin:5px 0 0 0; padding:0; line-height:20px;
}
#pagebody #left{
float:left; width:22%; height:350px; background-color:#CEE7FF; text-align:left;
}
#pagebody #left p{
        height:30px; line-height:30px; text-align: center; font-size:14px;
}
#pagebody #right{
        float:right; height:300px; width:75%; text-align:left;
}
#content{
        min-height:400px; min-width:600px;
}
#footer{
        clear:both; height:60px; width:100%; text-align:center; line-height:150%;
        font-family: Arial;
}
table.mytable1{
        width:98%; margin:0px auto; margin-top:15px; border-top:1px solid #0099CC;
        border-bottom:1px solid #0099CC; font-size:14px;
}
table.mytable1 th{
        background-color: #CEE7FF; border-left:1px dotted #0099CC; text-align: center;
        font-weight: normal;
}
table.mytable1 td{
        border-top:1px dotted #0099CC; border-left:1px dotted #0099CC; text-align:center;
}
table.mytable2{
        width:598px; margin:0px auto; margin-top:15px; border-top:1px solid #0099CC;
border-left:1px solid #0099CC; border-right:1px solid #0099CC;
```

```
}
table.mytable2 th{
    width:100px; background-color:#CEE7FF; border-bottom:1px solid #0099CC;
    border-right:1px solid #0099CC; padding-right:5px; text-align:right;
font-weight:normal;
}
table.mytable2 td{
    text-align: left;padding-left: 5px; border-bottom:1px solid #0099CC;
}
```

第 2 章　基于 Spring Boot 构建项目

本章主要介绍 Spring Boot，以及如何使用 IDEA 建立 Spring Boot 项目，并结合网上人才中心系统介绍如何进行实体类与接口设计，如何实现数据访问层、业务逻辑层、控制层和视图层。

2.1　Spring Boot 概述

2.1.1　Spring 及 Spring MVC

1. Spring

Spring 是一个开源框架，是为了解决企业级应用程序开发的复杂性而创建的。Spring 主要专注于如何利用类、对象和服务组成一个企业级应用，并通过规范的方式将各种不同的组件整合成一个完整的系统。它基于依赖注入和面向方面技术，大大地降低了应用开发的难度与复杂度，提高了开发的速度，为企业级应用提供了一个有效的、轻量级的解决方案。

Spring 是一个分层架构，由 7 个模块组成。Spring 的其他模块构建在 Spring Core 模块之上，该模块定义了创建、配置和管理 Bean 的方式，如图 2-1 所示。

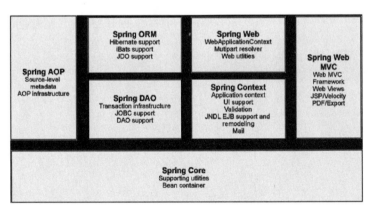

图 2-1　Spring 的 7 个模块

- Spring Core（核心容器）：该模块用于提供 Spring 的基本功能，其主要组件是 BeanFactory。BeanFactory 是工厂模式的实现，使用控制反转（IoC）模式将应用程序的配置和依赖性规范与实际的应用程序代码分开。

- Spring Context（上下文）：该模块建立在 Spring Core 模块的基础上，用于向 Spring 框架提供上下文信息。它提供了一种框架风格的方式来访问对象，有些类似于 JNDI

注册表。Context 封装包继承了 beans 包的功能，还增加了国际化（I18N）（用于规范 resource bundle）、事件传播、资源装载，以及透明创建上下文功能。
- Spring AOP（面向方面）：通过配置管理特性，Spring AOP 模块直接将面向方面的编程功能集成到了 Spring 中，可以很容易地使 Spring 管理的任何对象支持 AOP。Spring AOP 模块为基于 Spring 的应用程序中的对象提供了事务管理服务。通过使用 Spring AOP，不用依赖 EJB 组件，就可以将声明性事务管理集成到应用程序中。
- Spring DAO：该模块提供了 JDBC 的抽象层，可以消除冗长的 JDBC 编码和解析数据库厂商特有的错误代码。并且，JDBC 封装包提供了一种比编程性更好的声明性事务管理方法，不仅实现了特定接口，而且对所有的 POJO 都适用。
- Spring ORM：在 Spring 中插入了若干个 ORM 框架，从而提供了 ORM 的对象关系工具，包括 JDO、Hibernate 和 iBatis SQL Map。所有工具都遵从 Spring 的通用事务和 DAO 异常层次结构。
- Spring Web：该模块建立在 Spring Context 模块之上，为基于 Web 的应用程序提供了上下文。因此，Spring 支持与 Struts 的集成。Spring Web 模块还简化了处理大部分请求及将请求参数绑定到域对象的工作。
- Spring Web MVC：该模块提供了 Web 应用的 MVC 实现。Spring Web MVC（简称为 Spring MVC）并不仅仅提供一种传统的实现，它还在领域模型代码和 Web Form 之间提供了一种清晰的分离模型，并且可以借助于 Spring 的其他特性。

2．Spring MVC

Spring MVC 是 Spring 的一部分，属于 Spring 的 Web 模块，是一种基于 Java 实现了 Web MVC 设计模式的请求驱动类型的轻量级 Web 框架，也就是使用了 MVC 架构模式的思想，将 Web 层进行职责解耦的框架，其设计目的是帮助我们简化开发。

Spring MVC 使用标注来简化 Java Web 的开发，并且支持 RESTful 风格的 URL 请求。该框架具有松耦合、可插拔的组件结构，比其他的 MVC 框架更具扩展性和灵活性。

Spring MVC 是一个基于请求驱动的 Web 框架，并且使用了前端控制器模式来进行设计，可以根据请求映射规则将请求分发给相应的页面控制器（动作/处理器）进行处理。

1）Spring MVC 工作原理

（1）启动服务器，根据 web.xml 文件的配置加载前端控制器（也称总控制器）DispatcherServlet。在加载时，会完成一系列的初始化动作。

（2）根据 Servlet 的映射请求，并参照控制器配置文件（即 spmvc-servlet.xml 文件），把具体的请求分发给特定的后端控制器进行处理

（3）后端控制器调用相应的逻辑层代码，完成处理并返回视图对象（ModelAndView）给前端处理器。

（4）前端控制器根据后端控制器返回的 ModelAndView 对象，并结合一些配置（后面有说明），返回一个相应的页面给客户端。

2）Spring MVC 处理请求的流程

Spring MVC 处理请求的流程如图 2-2 所示。

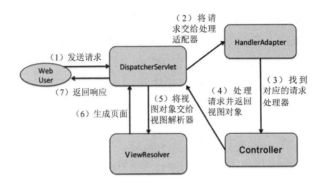

图 2-2 Spring MVC 处理请求的流程

（1）用户 Web User 发送请求至前端控制器 DispatcherServlet。

（2）DispatcherServlet 收到请求，将请求交给处理适配器 HandlerAdapter。

（3）HandlerAdapter 根据请求 URL 找到具体的处理器，生成处理器对象，并调用处理器 Controller（也叫后端控制器）。

（4）Controller 在处理完成后，返回视图对象 ModelAndView 给 DispatcherServlet。

（5）DispatcherServlet 将 ModelAndView 传给视图解析器 ViewResolver。

（6）ViewResolver 在解析后，返回具体的页面 View 给 DispatcherServlet。

（7）DispatcherServlet 对 View 进行渲染视图（即将模型数据填充至视图中），然后返回响应给 Web User。

2.1.2 Spring Boot

Spring Boot 是由 Pivotal 团队在 2013 年开始研发并于 2014 年 4 月发布第一个版本的开源的轻量级框架。它基于 Spring 4.0 设计，不仅继承了 Spring 原有的优秀特性，而且通过简化配置进一步简化了 Spring 应用的整个搭建和开发过程。另外，Spring Boot 通过集成大量的框架使得依赖包的版本冲突，以及引用的不稳定性等问题得到了很好的解决。

Spring Boot 引入自动配置的概念，使项目配置变得更容易。Spring Boot 本身并不提供 Spring 的核心特性及扩展功能，只是用来快速、敏捷地开发新一代基于 Spring 的应用程序。也就是说，它并不是用来替代 Spring 的解决方案，而是和 Spring 紧密结合以提升 Spring 开发者体验的工具。同时，它集成了大量常用的第三方库，如 Jackson、JDBC、Mongo、Redis、Mail 等。在 Spring Boot 应用中，这些第三方库几乎可以零配置地实现开箱即用（out-of-the-box），因此大部分的 Spring Boot 应用都只需要少量的配置代码，使得开发者能够更加专注于业务逻辑。也可以说，Spring Boot 是承载者，只是用来辅助开发者简化项目搭建过程的。

如果 Spring Boot 承载的是 Web 项目，并使用 Spring MVC 作为 MVC 框架，则其工作流程和 Spring MVC 的工作流程是完全一样的，因为这部分工作是 Spring MVC 做的而不是 Spring Boot 做的。此外，Web 应用需要使用 Tomcat 服务器启动，而 Spring Boot 内置服务器容器，只需通过@SpringBootApplication 注解类中的 main()函数即可启动。

Spring Boot 提供了 4 个核心功能，如下所述。

- 自动配置：针对很多 Spring 应用程序的常见应用功能，Spring Boot 能自动提供相关的配置。
- 起步依赖：用户只需告诉 Spring Boot 需要什么功能，它就能引入需要的库。
- Spring Boot CLI：用户只需编写代码就能完成完整的应用程序，无须构建传统项目。
- Actuator：让用户深入了解运行中的 Spring Boot 应用程序。

上述每一个特性都能以自己的方式简化 Spring 应用的开发。

2.2 使用 IDEA 创建 Spring Boot 项目

2.2.1 创建 Spring Boot 项目

创建 Spring Boot 项目的基本步骤如下所述。

（1）打开 IDEA，单击【Create New Project】按钮，开始创建一个新项目，如图 2-3 所示。如果已经打开了 IDEA，则选择【File】→【News】→【Project】命令，打开项目创建界面。

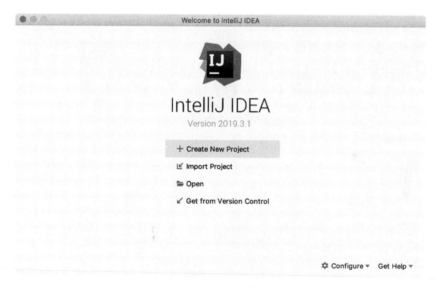

图 2-3　创建一个新项目

（2）如图 2-4 所示，在项目创建界面左侧列表框中选择【Spring Initializr】选项，然后单击【Next】按钮。注意，这里 IDEA 默认使用【https://start.spring.io】提供的在线模板，所以需要保证网络畅通。当然，也可以选中下面的【Custom】单选按钮，并从指定的链接加载模板。

Java EE 企业级应用开发技术研究

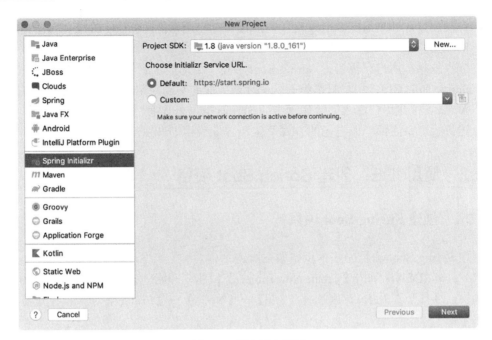

图 2-4　项目创建界面

（3）在如图 2-5 所示的界面中按实际情况依次填写项目信息。在【Type】下拉列表中可以选择【Maven Project】或【Pom】选项，这里保持默认设置，在【Packaging】下拉列表中选择【War】选项，并分别设置【Group】【Artifact】【Package】，然后单击【Next】按钮。

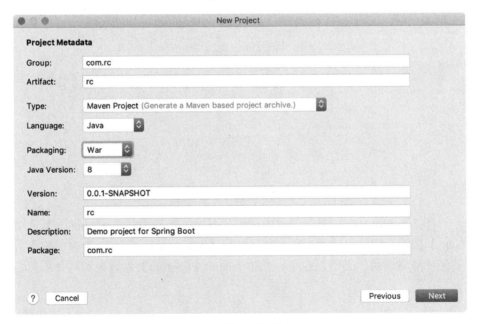

图 2-5　填写项目信息

（4）在如图 2-6 所示的界面中选择依赖库。首先从【Spring Boot】下拉列表中选择 Spring

Boot 的版本，此处保持默认设置。然后在界面左侧列表框中选择大类，并在界面中间勾选需要的依赖所对应的复选框，这里选择了【Developer Tools】类别下的【Spring Boot DevTools】【Lombok】【Spring Configuration Processor】，【Web】类别下的【Spring Web】，以及【SQL】类别下的【JDBC API】和【MySQL Driver】。最后单击【Next】按钮。

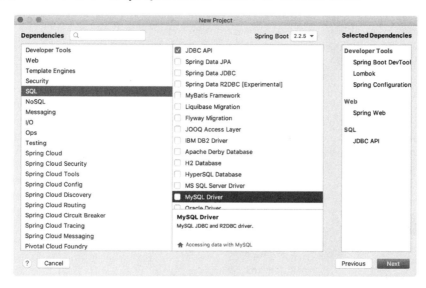

图 2-6　选择依赖库

（5）在如图 2-7 所示的界面中设置项目名称和存储位置，并在完成后单击【Finish】按钮。

图 2-7　设置项目名称和存储位置

（6）在使用 IDEA 构建好项目后，项目结构如图 2-8 所示。

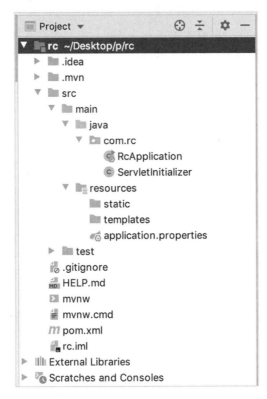

图 2-8　项目结构

2.2.2　根据项目需要引入其他依赖

1．初始依赖说明

对于刚刚创建的项目而言，pom.xml 文件的内容如下：

```
<?xml version="1.0" encoding="UTF-8"?>
<project xmlns="http://maven.apache.org/POM/4.0.0"
         xmlns:xsi="http://www.w3.org/2001/XMLSchema-instance"
         xsi:schemaLocation="http://maven.apache.org/POM/4.0.0
         https://maven.apache.org/xsd/maven-4.0.0.xsd">
    <modelVersion>4.0.0</modelVersion>
    <parent>
        <groupId>org.springframework.boot</groupId>
        <artifactId>spring-boot-starter-parent</artifactId>
        <version>2.2.5.RELEASE</version>
        <relativePath/> <!-- lookup parent from repository -->
    </parent>
    <groupId>com.rc</groupId>
    <artifactId>rc</artifactId>
    <version>0.0.1-SNAPSHOT</version>
```

```xml
<packaging>war</packaging>
<name>rc</name>
<description>Demo project for Spring Boot</description>
<properties>
    <java.version>1.8</java.version>
</properties>
<dependencies>
    <dependency>
        <groupId>org.springframework.boot</groupId>
        <artifactId>spring-boot-starter-jdbc</artifactId>
    </dependency>
    <dependency>
        <groupId>org.springframework.boot</groupId>
        <artifactId>spring-boot-starter-web</artifactId>
    </dependency>
    <dependency>
        <groupId>org.springframework.boot</groupId>
        <artifactId>spring-boot-devtools</artifactId>
        <scope>runtime</scope>
        <optional>true</optional>
    </dependency>
    <dependency>
        <groupId>mysql</groupId>
        <artifactId>mysql-connector-java</artifactId>
        <scope>runtime</scope>
    </dependency>
    <dependency>
        <groupId>org.springframework.boot</groupId>
        <artifactId>spring-boot-configuration-processor</artifactId>
        <optional>true</optional>
    </dependency>
    <dependency>
        <groupId>org.projectlombok</groupId>
        <artifactId>lombok</artifactId>
        <optional>true</optional>
    </dependency>
    <dependency>
        <groupId>org.springframework.boot</groupId>
        <artifactId>spring-boot-starter-tomcat</artifactId>
        <scope>provided</scope>
    </dependency>
    <dependency>
        <groupId>org.springframework.boot</groupId>
        <artifactId>spring-boot-starter-test</artifactId>
        <scope>test</scope>
        <exclusions>
            <exclusion>
                <groupId>org.junit.vintage</groupId>
                <artifactId>junit-vintage-engine</artifactId>
```

```xml
                    </exclusion>
                </exclusions>
            </dependency>
        </dependencies>
        <build>
            <plugins>
                <plugin>
                    <groupId>org.springframework.boot</groupId>
                    <artifactId>spring-boot-maven-plugin</artifactId>
                </plugin>
            </plugins>
        </build>
</project>
```

pom.xml 文件中包含的依赖说明如下所述。

- spring-boot-starter-parent：作为父依赖，提供 Spring Boot 的默认配置和一棵完整的 Spring Boot 依赖树，还包括 Maven 的打包插件。它可以管理 Spring Boot 需要的依赖，从而统一各种 jar 包的版本号，避免了因版本不一致而出现的问题。所以，引入其他的依赖就可以省略版本号。当然也可以加上指定的版本号，从而取代默认的版本号。
- spring-boot-starter-jdbc：使用 JDBC 操作数据库的相关 jar 包。它可以根据配置自动创建数据源和 JdbcTemplate 对象。在程序中，可以通过注入 JdbcTemplate 对象来操作数据库，也可以在 application.properties 文件中配置数据源。
- spring-boot-starter-web：自动引入了 Web 模块开发需要的相关 jar 包。默认使用嵌套式的 Tomcat 作为 Web 容器对外提供 HTTP 服务，默认端口 8080 对外提供监听服务。若想改变默认的配置端口，则可以在 application.properties 文件中指定。
- spring-boot-devtools：支持热部署，提高开发者的开发效率。
- mysql-connector-java：MySQL 驱动程序。
- spring-boot-configuration-processor：配置文件 xml 或 properties、yml 的处理器。
- lombok：提供了一组有用的标注，用来消除 Java 类中的大量样板代码，使得 Java 类更加简洁且易于维护。例如，@Data 可生成 setter/getter、equals、canEqual、hashCode、toString 方法；@Slf4j 注解在类上，可生成 log 变量。在 IDEA 编写源码时，需要先安装 Lombok 插件。
- spring-boot-starter-tomcat：在部署 war 包时，需要依赖此包。

2. 引入其他依赖

修改 pom.xml 文件，引入依赖，以支持 JSP，修改内容如下：

```xml
<dependency>
    <groupId>org.apache.tomcat.embed</groupId>
    <artifactId>tomcat-embed-jasper</artifactId>
    <scope>provided</scope>
</dependency>
<dependency>
    <groupId>junit</groupId>
```

```
<artifactId>junit</artifactId>
<scope>test</scope>
</dependency>
```

2.2.3　按分层结构组织程序结构

按分层结构组织程序，还需要建立其他包，这些包建立在 com.rc 包下，主要包括 com.rc.controller（控制层包）、com.rc.dao（数据访问层接口包）、com.rc.dao.impl（数据访问层实现类包）、com.rc.service（业务逻辑层接口包）、com.rc.service.impl（业务逻辑层实现类包）、com.rc.entity（实体类包），结构如图 2-9 所示。

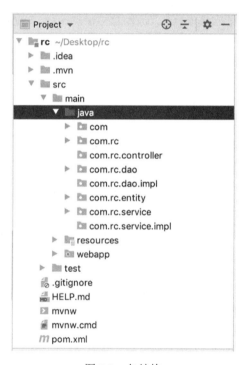

图 2-9　包结构

Spring Boot 默认使用的是 Thymeleaf 模板引擎，HTML 网页存放在/resources/templates 文件夹下，静态资源存放在 resources/static 文件夹下。在使用 JSP 时，不再使用这两个文件夹，而使用 webapp 文件夹。在 webapp 文件夹下创建 WEB-INF 文件夹，并在 WEB-INF 文件夹下创建 jsp 文件夹用于存放 JSP 网页，同时在 webapp 文件夹下创建 static 文件夹用于存放静态资源。

视图层的相关文件存放在 WEB-INF/jsp 文件夹下，在 jsp 文件夹下建立 user（用户）、role（角色）、permission（权限）、application（应聘申请）、common（公共）、company（企业）、job（职位）、news（新闻）、person（个人）、upfiles（上传文件）文件夹，分别用于存放不同的网页文件。

在 static 文件夹下建立 js（JavaScript）、css（样式表）、images（图像）文件夹，分别用

于存放静态资源。

另外，在 jsp 文件夹中创建 index.jsp 作为系统首页，创建 manage.jsp 作为管理页。为了测试，可以在内容中先输入 OK。视图层文件夹结构如图 2-10 所示。

图 2-10　视图层文件夹结构

com.rc 包下的两个类的说明如下所述。

RcApplication 类中包含一个 main()方法，是程序的启动类。它使用内置的 Tomcat 来部署程序。@SpringBootApplication 标注用于完成自动配置。例如：

```
@SpringBootApplication
public class RcApplication {
    public static void main(String[] args) {
        SpringApplication.run(RcApplication.class, args);
    }
}
```

ServletInitializer 类使用外部 Tomcat 来部署程序。在建立项目时，若选择打成 war 包，则会自动创建这个类。该类继承于 SpringBootServletInitializer 并实现了 configure()方法，与原来的启动类配合，可完成自动配置。例如：

```
public class ServletInitializer extends SpringBootServletInitializer {
    @Override
    protected SpringApplicationBuilder configure(SpringApplicationBuilder application) {
        return application.sources(RcApplication.class);
    }
}
```

2.2.4 建立分页工具类

建立一个分页工具类 PageHelper，代码如下：

```java
package com.rc.util;
/**
 * @description:分页工具类
 * @className: PageHelper
 * @author: Yang ShuLin
 * @create: 2020-02-23  13:49
 */
public class PageHelper {
    private String pageBar;
    private String numPageBar;
    public PageHelper(Long count, Integer pageNo, Integer pageSize) {
        StringBuffer sb = new StringBuffer();
        if (count == 0) {
            sb.append("<script>");
            sb.append(" function goPage(n){"};
            sb.append("document.getElementsByName('pageNo')[0].value=n;");
            sb.append("document.forms[0].submit()");
            sb.append(")");
            sb.append("</script>");
            pageBar = sb.toString();
            numPageBar = pageBar;
            return;
        }
        //计算总页数
        int pageCount = (int) (count / pageSize + (count % pageSize == 0 ? 0 : 1));
        //判断当前页号的合法性
        if (pageNo < 1) {
            pageNo = 1;
        }
        if (pageNo > pageCount) {
            pageNo = pageCount;
        }
        sb = new StringBuffer();
        sb.append("每页:").append(pageSize).append("  页次:").append(pageNo).append
("/").append(pageCount).append("  总计:").append(count).append("  ");
        if (pageNo == 1) {
            sb.append("<span class=\"page1\">首页</span><span class=\"page1\">上页</span>");
        } else {
            sb.append("<a class=\"page\" href=\"JavaScript:goPage(1);\">首页</a> ");
            sb.append("<a class=\"page\" href=\"JavaScript:goPage(").append(pageNo - 1).append(");
\">上页</a>");
        }
        if (pageNo == pageCount) {
            sb.append("<span class=\"page1\">下页</span><span class=\"page1\">尾页</span>");
        } else {
```

```
        sb.append("<a class=\"page\" href=\"JavaScript:goPage(").append(pageNo + 1).append(");
\">下页</a>");
        sb.append("<a class=\"page\" href=\"JavaScript:goPage(").append(pageCount).append(");
\">尾页</a>");
}
sb.append("<script>");
sb.append(" function goPage(n){");
sb.append("document.getElementsByName('pageNo')[0].value=n;");
sb.append("document.forms[0].submit()");
sb.append(")");
sb.append("</script>");
pageBar = sb.toString();
sb = new StringBuffer();
int start;
int end;
if (pageNo > 5) {
    end = pageNo+4;
    if(end > count.intValue()) end = count.intValue();
    start=end-9;
    if(start<1)start=1;
}else{
    start=1;
    end=start+9;
    if(end>count.intValue()){
        end=count.intValue();
    }
}
if (pageNo > 1) {
    sb.append("<a class=\"numpage prev\" href=\"JavaScript:goPage(").append(pageNo - 1).
append(");\">前一页</a>");
}
for (int i = start; i <= end; i++) {
    if (pageNo != i) {
        sb.append("<a class=\"numpage\" href=\"JavaScript:goPage(").append(i).append(");
\">").append(i).append("</a>");
    } else {
        sb.append("<span class=\"numpage1\">").append(i).append("</span>");
    }
}
if (pageNo < pageCount) {
    sb.append("<a class=\"numpage next\" href=\"JavaScript:goPage(").append(pageNo + 1).
append(");\">下一页</a>");
}
sb.append("<script>");
sb.append(" function goPage(n){");
sb.append("document.getElementsByName('pageNo')[0].value=n;");
sb.append("document.forms[0].submit()");
sb.append("}");
sb.append("</script>");
```

```java
        numPageBar = sb.toString();
    }
    public void setPageBar(String pageBar) {
        this.pageBar = pageBar;
    }
    public void setNumPageBar(String numPageBar) {
        this.numPageBar = numPageBar;
    }
    public String getNumPageBar() {
        return numPageBar;
    }
    public String getPageBar() {
        return pageBar;
    }
}
```

2.2.5　应用程序基本配置

在 application.properties 文件中添加如下配置：

```
#配置程序端口，默认为 8080
server.port=8080
#用户会话 session 的过期时间
server.servlet.session.timeout=30
#配置默认访问路径，默认为/
server.servlet.context-path=/rc
#配置 Tomcat 编码，默认为 UTF-8
server.tomcat.uri-encoding=UTF-8
#配置最大线程数
server.tomcat.max-threads=1000
#数据源配置
spring.datasource.url=jdbc:mysql://localhost:3306/rc?useUnicode=true&characterEncoding=utf8&serverTimezone=GMT&useSSL=false&zeroDateTimeBehavior=convertToNull&allowMultiQueries=true
spring.datasource.username=root
spring.datasource.password=12345
spring.datasource.driverClassName=com.mysql.cj.jdbc.Driver
#视图解析配置
spring.mvc.view.prefix=/WEB-INF/pages/
spring.mvc.view.suffix=.jsp
```

2.3　实体类与接口设计

2.3.1　实体类设计

为了简化代码，在进行实体类设计时使用@Data 标注。因篇幅限制，这里仅给出两个实体类，代码如下：

```java
package com.rc.entity;
```

```java
import lombok.Data;
import java.util.Date;
/**
 * 名称: RcUser
 * 描述: 用户实体类
 * @version 1.0
 * @author: Yang ShuLin
 * @datetime: 2020-03-07 19:00
 * @version 1.0
 */
@Data
public class RcUser {
    private Integer userId;
    private String userName;
    private String userRealname;
    private String userPwd;
    private String userType;
    private String userEmail;
    private Date userPosttime;
    private RcCompany company;
}

package com.rc.entity;
import lombok.Data;
/**
 * 名称: RcUser
 * 描述: 企业实体类
 * @version 1.0
 * @author: Yang ShuLin
 * @datetime: 2020-03-07 19:00
 * @version 1.0
 */
@Data
public class RcCompany {
    private String comId;
    private String comName;
    private String comType;
    private String comCreatetime;
    private String comContactor;
    private Integer comEmpnum;
    private String comTel;
    private String comFax;
    private String comAddr;
    private String comZip;
    private String comHomepage;
    private String comEmail;
    private String comDesc;
    private Date comPosttime;
}
```

2.3.2 业务逻辑层接口设计

业务逻辑层接口体现了业务功能的需要,反映了系统所具有的业务功能,均被存放在 com.rc.service 包下。以企业业务逻辑接口为例,代码如下:

```
package com.rc.service;
import java.util.List;
import com.rc.entity.RcCompany;
public interface RcCompanyService {
    void addCompany(RcCompany com);
    void editCompany(RcCompany com);
    void deleteCompany(Integer comId);
    RcCompany findCompanyById(Integer comId);
    List<RcCompany> findCompanies(Integer num);
    List<RcCompany> findCompanies(String comType, String comName, Integer pageNo, Integer pageSize);
    Integer findCount(String comType, String comName);
}
```

2.3.3 数据访问层接口设计

数据访问层接口是根据业务功能的需要进行设计的,反映了系统为实现业务功能所要完成的数据库操作,均被存放在 com.rc.dao 包下。以企业数据访问接口为例,代码如下:

```
package com.rc.dao;
import java.util.List;
import com.rc.entity.RcCompany;
public interface RcCompanyDao{
    void insert(RcCompany com);
    void update(RcCompany com);
    void delete(Integer userId);
    RcCompany selectById(Integer comId);
    List<RcCompany> selectSome(Integer num);
    List<RcCompany> selectSomeByPage(String comType, String comName, Integer pageNo, Integer pageSize);
    Integer selectCount(String comType, String comName);
}
```

2.4 数据访问层与业务逻辑层实现

2.4.1 数据访问层实现

数据访问层的类通过@Resource 注入 JdbcTemplate 对象来执行 SQL 语句。类本身是通过@Repository 标注的,以表明它是数据访问组件。这样,容器就可以根据业务逻辑层的需要注入数据访问层对象。

以 RcCompanyDaoImpl 类为例，代码如下：

```java
package com.rc.dao.impl;
import com.rc.dao.RcCompanyDao;
import com.rc.entity.RcCompany;
import org.springframework.jdbc.core.BeanPropertyRowMapper;
import org.springframework.jdbc.core.JdbcTemplate;
import org.springframework.stereotype.Repository;
import javax.annotation.Resource;
import java.util.List;
/**
 * 名称：RcCompanyDaoImpl
 * 描述：企业数据访问实现类
 * @author: Yang ShuLin
 * @datetime: 2020-03-07 18:13
 * @version 1.0
 */
@Repository
public class RcCompanyDaoImpl implements RcCompanyDao {
    @Resource
    private JdbcTemplate jdbcTemplate;

    @Override
    public List<RcCompany> selectSome(Integer num) {
        List<RcCompany> list = jdbcTemplate.query("select * from rc_company order by com_postTime desc limit 0,?", new Object[]{num}, new BeanPropertyRowMapper(RcCompany.class));
        return list;
    }

    @Override
    public List<RcCompany> selectSomeByPage(String comType, String comName, Integer pageNo, Integer pageSize) {
        List<RcCompany> list = jdbcTemplate.query("select * from rc_company where com_type like ? and com_name like ? order by com_postTime desc limit ?,?", new Object[]{"%"+comType+"%","%"+comName+"%",(pageNo-1)*pageSize,pageSize}, BeanPropertyRowMapper.newInstance(RcCompany.class));
        return list;
    }

    @Override
    public Integer selectCount(String compType, String comName) {
        Integer count=jdbcTemplate.queryForObject("select count(*) from rc_company where com_type like ? and com_name like ?", new Object[]{"%"+compType+"%","%"+comName+"%"}, Integer.class);
        return count;
    }

    @Override
    public void insert(RcCompany company) {
        jdbcTemplate.update("insert into rc_company(com_name,com_type,com_createTime,com_contactor,
```

```
com_empNum,com_tel,com_fax,com_addr,com_zip,com_homepage,com_email,com_desc) values
(?,?,?,?,?,?,?,?,?,?,?,?,?,?,?)",company.getComName(),company.getComType(),company.getComCreateTime(),
company.getComContactor(),company.getComEmpNum(),company.getComTel(),company.getComFax(),
company.getComAddr(),company.getComZip(),company.getComHomepage(),company.getComEmail(),
company.getComDesc());
}

@Override
public void update(RcCompany company) {
    jdbcTemplate.update("update rc_company set user_name=?,user_pwd=?,user_type=?,com_create
Time=?,com_name=?,com_type=?,com_contactor=?,com_empNum=?,com_tel=?,com_fax=?,com_
addr=?,com_zip=?,com_homepage=?,com_email=?,com_desc=? where user_id = ?", company.get
UserName(),company.getUserPwd(),company.getUserType(),company.getComCreateTime(),company.
getComName(),company.getComType(),company.getComContactor(),company.getComEmpNum(),
company.getComTel(),company.getComFax(),company.getComAddr(),company.getComZip(),company.
getComHomepage(),company.getComEmail(),company.getComDesc(),company.getComId());
}

@Override
public void delete(Integer comId) {
    jdbcTemplate.update("delete from rc_company where com_id = ?",comId);
}

@Override
public RcCompany selectById(Integer comId) {
    RcCompany company = jdbcTemplate.queryForObject("select * from rc_company where
com_id = ?", new Object[]{userId}, BeanPropertyRowMapper.newInstance(RcCompany.class));
    return company;
}
}
```

上面的类并没有涉及关联查询，如果涉及关联查询，则需要使用 RowMapper。例如，在职位查询中，按 num 查询新的职位，代码如下：

```
@Override
public List<RcJob> selectSome(Integer num) {
    List<RcJob> list = jdbcTemplate.query("select a.*,b.com_name from rc_job a,rc_company b where
a.com_id=b.user_id order by job_posttime desc limit 0,?", new Object[]{num}, new RowMapper<RcJ
ob>() {
        @Override
        public RcJob mapRow(ResultSet rs, int i) throws SQLException {
            RcJob job = new RcJob();
            job.setJobId(rs.getInt("job_id"));
            job.setJobTitle(rs.getString("job_title"));
            job.setJobType(rs.getString("job_type"));
            job.setJobEducation(rs.getString("job_education"));
            job.setJobSex(rs.getString("job_sex"));
            job.setJobAge(rs.getString("job_age"));
            job.setJobWorktime(rs.getString("job_worktime"));
            job.setJobWorkplace(rs.getString("job_workplace"));
```

```
                job.setJobNum(rs.getInt("job_num"));
                job.setJobEndtime(rs.getString("job_endtime"));
                job.setJobComment(rs.getString("job_comment"));
                job.setJobPosttime(rs.getTimestamp("job_posttime"));
                RcCompany c=new RcCompany();
                c.setComId(rs.getInt("com_id"));
                c.setComName(rs.getString("com_name"));
                job.setCompany(c);
                return job;
            }
        });
        return list;
}
```

2.4.2 对数据访问层进行单元测试

（1）在 RcCompanyDaoImpl 类源码的名称上单击鼠标右键，并在弹出的快捷菜单中选择【Go To】→【Test】→【Create New Test】命令，打开【Create Test】对话框，创建测试类，如图 2-11 所示，并单击【OK】按钮。

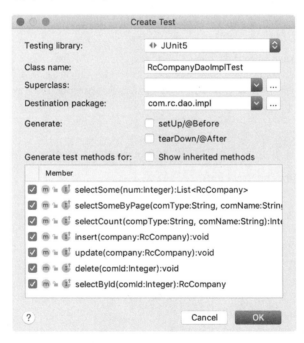

图 2-11　创建测试类

（2）完善创建的测试类，代码如下：

```
@RunWith(SpringRunner.class)
@SpringBootTest
@Slf4j
class RcCompanyDaoImplTest {
```

```java
@Resource
private RcCompanyDao companyDao;

@Test
void selectSomeByPage() {
    try {
        List<RcCompany> list=companyDao.selectSomeByPage("","",1,2);
        log.info(list.toString());
    }catch (Exception e){
        e.printStackTrace();
    }
}

@Test
void testSelectSome() {
    try {
        List<RcCompany> list=companyDao.selectSome(3);
        log.info(list.toString());
    }catch (Exception e){
        e.printStackTrace();
    }
}

@Test
void selectCount() {
    try {
        Integer count=companyDao.selectCount("","");
        log.info(count+"");
    }catch (Exception e){
        e.printStackTrace();
    }
}

@Test
void insert() {
    RcCompany company=new RcCompany();
    company.setComName("ddssd");
    company.setComType("dsd");
    company.setComCreatetime("2015-02-12");
    company.setComContactor("zhang");
    company.setComAddr("北京");
    company.setComEmail("dd@ssc.com");
    company.setComTel("34343443");
    company.setComFax("3332");
    company.setComZip("343434");
    company.setComEmpnum(50);
    company.setComHomepage("http://dfdsdsf");
    company.setComDesc("dssddsds");
    try {
```

```
            companyDao.insert(company);
        }catch (Exception e){
            e.printStackTrace();
        }
        //company.setComPostTime();
    }

    @Test
    void update() {
        RcCompany company=new RcCompany();
        company.setComId(1);
        company.setComName("ddssd");
        company.setComType("dsd");
        company.setComCreatetime("2015-02-12");
        company.setComContactor("zhang");
        company.setComAddr("北京");
        company.setComEmail("dd@ssc.com");
        company.setComTel("34343443");
        company.setComFax("3332");
        company.setComZip("343434");
        company.setComEmpnum(50);
        company.setComHomepage("http://dfdsdsf");
        company.setComDesc("dssddsds");
        try {
            companyDao.update(company);
        }catch (Exception e){
            e.printStackTrace();
        }
    }

    @Test
    void delete() {
    }
    @Test
    void selectById() {
        try {
            RcCompany company=companyDao.selectById(1);
            log.info(company.toString());
        }catch (Exception e){
            e.printStackTrace();
        }
    }
}
```

（3）在 RcCompanyDaoImpl 类源码的某个方法的名称上单击鼠标右键，并在弹出的快捷菜单中选择运行该方法的命令。

（4）查看运行效果。

2.4.3 业务逻辑层实现

业务逻辑层的类通过@Resource 注入数据访问层对象。类本身是通过@Service 标注的，以表明它是业务逻辑组件。这样，容器就可以根据控制层的需要注入业务逻辑层对象。

以 RcCompanyServiceImpl 类为例，代码如下：

```java
package com.rc.service.impl;
//导包语句省略
/**
 * 名称：RcCompanyServiceImpl
 * 描述：企业业务逻辑实现类
 * @author: Yang ShuLin
 * @datetime: 2020-03-07 21:33
 * @version 1.0
 */
@Service
public class RcCompanyServiceImpl implements RcCompanyService {
    @Resource
    private RcCompanyDao companyDao;

    @Override
    public void addCompany(RcCompany com) {
        companyDao.insert(com);
    }

    @Override
    public void editCompany(RcCompany com) {
        companyDao.update(com);
    }

    @Override
    public void deleteCompany(Integer comId) {
        companyDao.delete(comId);
    }

    @Override
    public RcCompany findCompanyById(Integer comId) {
        return companyDao.selectById(comId);
    }

    @Override
    public List<RcCompany> findCompanies(Integer num) {
        return companyDao.selectSome(num);
    }

    @Override
    public List<RcCompany> findCompanies(String comType, String comName, Integer pageNo, Integer pageSize) {
        return companyDao.selectSomeByPage(comType,comName,pageNo,pageSize);
```

```
    }
    @Override
    public Integer findCount(String comType, String comName) {
        return companyDao.selectCount(comType,comName);
    }
}
```

2.5 控制层实现

控制层用于处理 HTTP 请求。基于 Spring Boot 的控制层设计与基于 Spring MVC 的控制层设计类似。

2.5.1 控制层设计的基本原理

1. 基本标注

控制层的类使用@Controller 标注或@RestController 标注，两者的区别如下所述。

对于使用@Controller 标注的类，若它的方法使用@RequestMapping、@GetMapping 或@PostMapping 标注，则返回值对应的是一个视图；若它的方法使用@ResponseBody 标注，则返回 JSON 数据。而对于使用@RestController 标注的类，其方法均返回 JSON 数据。

@RequestMapping 是一个用来处理请求地址映射的标注，可用于类或方法上。当将其用于类上时，表示该类中的所有响应请求的方法都是以该地址为父路径的；当将其用于方法上时，表示请求该方法的路径。例如：

```
@Controller
@RequestMapping("/job")
public class RcJobController{
    …
    @RequestMapping("/add")
    public ModelAndView add(RcJob job){
        …
        return new ModelAndView(…)
    }
}
```

按照上面的定义，请求 add()方法的路径为/job/add。@RequestMapping 默认可以接收 POST 请求或 GET 请求。若需要明确请求的 method 类型，则可以使用 value 指定请求的路径，并使用 method 指定请求的 method 类型。例如：

```
@RequestMapping(value="/add", method="POST")
```

此外，也可以使用@GetMapping、@PostMapping 标注来直接明确请求的 method 类型。

2. 方法返回值的类型

对于使用@Controller 标注的类，其方法返回的值有以下几种情况。

1）返回 ModelAndView
- new ModelAndView("viewPath")：转发到视图层。
- new ModelAndView("viewPath","data",data)：转发到视图层，并携带数据 data。
- new ModelAndView("redirect:controllerPath")；重定向到控制层。

2）返回字符串
- return " viewPath"：转发到视图层。
- return "forward: viewPath"：转发到视图层。
- return "redirect: controllerPath"：重定向到控制层。

3）返回 JSON 数据

在方法上使用@ResponseBody 标注时，该方法返回值为 JSON 数据。

3．视图解析

Spring Boot 视图层默认使用 Thymeleaf 模板，视图解析的配置如下：

```
spring.thymeleaf.prefix=classpath:/templates/
spring.thymeleaf.suffix=.html
```

若 viewPath 为 pages/job/add，则实际指向的是/templates/pages/job/add.html。

如果视图层使用 JSP，需要引用如下类库：

```xml
<!--用于编译 jsp-->
<dependency>
    <groupId>org.apache.tomcat.embed</groupId>
    <artifactId>tomcat-embed-jasper</artifactId>
    <scope>provided</scope>
</dependency>
```

此外，在 main 文件夹下创建 webapp 文件夹，并在 webapp 文件夹下创建 WEB-INF 文件夹，并将视图层的 JSP 网页存放在该路径下。这时，视图解析的配置如下：

```
spring.mvc.view.prefix=/WEB-INF/pages/
spring.mvc.view.suffix=.jsp
```

在这种情况下，如果 viewPath 为 pages/job/add，则实际指向的是/WEB-INF/pages/job/add.jsp。

4．方法的参数

1）直接绑定方式

对于简单类型的参数（指 8 种基本类型的数据及它们的包装类，还有 String 类型），请求参数的 key 和 controller()方法中的形参名称一致。例如：

```java
@RequestMapping("/paraBind1")
public String paraBind1(Integer id){
    log.info("简单参数直接绑定，获取的参数 id："+id);
    return "hello";
}
```

2）使用@RequestParam 绑定

当名称不一致时，使用@RequestParam 标注。例如：

```java
@RequestMapping("/paraBind2")
public String paraBinding2(@RequestParam("item") Integer id){
```

```
    log.info("简单参数通过标注绑定，获取的参数 id："+id);
    return "hello";
}
```

3）对象类型的参数

当参数为实体对象时，要求请求中的参数和对象类中的属性名称保持一致。例如：

```
@RequestMapping("/paraBind3")
public String paraBinding3(User user){
    log.info("pojo 参数绑定，获取的 user："+user);
    return "hello";
}
```

4）JSON 格式参数

HTTP 请求体的内容是一个 JSON 数据，使用@RequestBody 标注。例如：

```
@PostMapping("/paraBind4")
public String paraBind4(@RequestBody User user) {
    log.info("JSON 参数绑定，获取的 User："+user);
    return "hello";
}
```

5）获取请求头

通过@RequestHeader 标注参数，可以获得头信息。例如：

```
@GetMapping("/paraBind5")
public String paraBind5(@RequestHeader String name) {
    log.info("请求头："+name);
    return "hello";
}
```

6）获取路径参数

通过@PathVariable 标注，可以获取 URL 中拼接的一个值。例如：

```
@GetMapping("/paraBind6/{name}")
public String paraBind6(@PathVariable("name") String name) {
    log.info("获取 URL 中拼接的一个值："+name);
    return "hello";
}
```

此外，可以直接使用 HttpServletRequest 或 HttpServletResponse 类型的参数，获取请求和响应对象。

2.5.2 控制类基类设计

创建一个控制类基类。该类定义了一个方法，用于进行统一异常处理。该方法使用了@ExceptionHandler 标注。代码如下：

```
package com.rc.controller;
import lombok.extern.slf4j.Slf4j;
import org.springframework.web.bind.annotation.ExceptionHandler;
import org.springframework.web.bind.annotation.ResponseBody;
import org.springframework.web.servlet.ModelAndView;
/**
 * 名称：RcBaseController
```

```
 * 描述：控制类基类
 * @author: Yang ShuLin
 * @datetime: 2020-03-07 15:53
 * @version 1.0
 */
@Slf4j
public class RcBaseController {
    @ExceptionHandler(value = {Exception.class})
    @ResponseBody
    public ModelAndView ExceptionHandler(Exception e) {
        log.debug("出现错误：",e.getMessage());
        return new ModelAndView("common/message","msg","出现错误，操作失败！");
    }
}
```

2.5.3 实现其他控制类

其他控制类继承于 RcBaseController 类。例如，企业控制类代码如下：

```
package com.rc.controller;
//导包语句省略
/**
 * 名称: RcCompanyController
 * 描述：企业控制类
 * @author: Yang ShuLin
 * @datetime: 2020-03-07 21:38
 * @version 1.0
 */
@Controller
@RequestMapping("/company")
public class RcCompanyController extends RcBaseController {
    @Resource
    private RcCompanyService companyService;

    @RequestMapping("/willAdd")
    public String willAdd() {
        return "company/add";
    }

    @RequestMapping("/add")
    public String add(RcCompany company) {
        companyService.addCompany(company);
        return "redirect:/company/manage";
    }

    //管理员要修改企业
    @RequestMapping("/willEdit/{id}")
    public ModelAndView willEdit(@PathVariable("id") Integer userId) {
        RcCompany company = companyService.findCompanyById(userId);
```

```java
        return new ModelAndView("company/edit", "company", company);
    }

    @RequestMapping("/edit")
    public String edit(RcCompany company) {
        companyService.editCompany(company);
        return "redirect:/company/manage";
    }

    @RequestMapping("/delete/{id}")
    public String delete(@PathVariable("id") Integer userId) {
        companyService.deleteCompany(userId);
        return "redirect:/company/manage";
    }

    @RequestMapping("/show/{id}")
    //@CrossOrigin(origins = "*")
    public ModelAndView show(@PathVariable("id") Integer userId) {
        RcCompany company = companyService.findCompanyById(userId);
        return new ModelAndView("company/show","company",company);
    }

    @RequestMapping("/manage")
    public ModelAndView findCompanies(String comType, String comName, Integer pageNo, Integer pageSize) {
        ModelMap map = new ModelMap();
        if(comType==null)comType="";
        if(comName==null)comName="";
        if(pageNo==null)pageNo=1;
        if(pageSize==null)pageSize=10;
        List<RcCompany> list = companyService.findCompanies(comType, comName, pageNo, pageSize);
        Integer count = companyService.findCount(comType, comName);
        PageHelper pageHelper = new PageHelper(Long.valueOf(count), pageNo, pageSize);
        map.put("comType", comType);
        map.put("comName", comName);
        map.put("pageNo", pageNo);
        map.put("pageSize", pageSize);
        map.put("list", list);
        map.put("pageHelper ", pageHelper);
        return new ModelAndView("company/manage", "map", map);
    }

    @RequestMapping("/browse")
    @ResponseBody
    public ModelAndView browse(String comType, String comName, Integer pageNo, Integer pageSize) {
        ModelMap map = new ModelMap();
        if(comType==null)comType="";
        if(comName==null)comName="";
        List<RcCompany> list = companyService.findCompanies(comType, comName, pageNo, pageSize);
        Integer count = companyService.findCount(comType, comName);
```

```
            PageHelper pageHelper = new PageHelper(Long.valueOf(count), pageNo, pageSize);
            map.put("comType", comType);
            map.put("comName", comName);
            map.put("pageNo", pageNo);
            map.put("pageSize", pageSize);
            map.put("list", list);
            map.put("pageHelper ", pageHelper);
            return new ModelAndView("company/browse", "map", map);
    }
}
```

2.5.4 对控制层进行单元测试

1．MockMvc

MockMvc 是由 spring-test 包提供的，实现了对 HTTP 请求的模拟，能够直接使用网络的形式调用 Controller，使得测试速度快、不依赖于网络环境。同时，MockMvc 提供了一套验证的工具，使得结果的验证非常方便。

MockMvcBuilder 接口提供了一个唯一的 build()方法用来构造 MockMvc，主要包括 StandaloneMockMvcBuilder 和 DefaultMockMvcBuilder 两个实现，分别对应两种测试方式，即独立安装和集成 Web 环境测试（并不会集成真正的 Web 环境，而是会通过相应的 Mock API 进行模拟测试，无须启动服务器）。同时 MockMvcBuilder 接口提供了对应的创建方法 standaloneSetup()和 webAppContextSetup()，在使用时直接调用即可。

2．对控制层进行测试

在下面的测试类中，注入了 MockMvc 对象，定义了一个工具性方法 doAction()，其他测试方法调用这个方法。注意，测试方法仅给出对分页查询的测试。代码如下：

```
package com.rc.controller;
//导包语句省略
@RunWith(SpringRunner.class)
@SpringBootTest
@AutoConfigureMockMvc
@Slf4j
class RcCompanyControllerTest {
    @Autowired
    private MockMvc mockMvc;

    //使用@LocalServerPort 将端口注入
    @Value("${server.port}")
    private int port;

    private void doAction(String method, String action, MultiValueMap params, MultiValueMap queryParams) {
        //构建请求
        MockHttpServletRequestBuilder request;
        if(method.toUpperCase().equals("POST")) {
            if(params==null)params=new LinkedMultiValueMap<>();
```

```
            request = MockMvcRequestBuilders.post("http://localhost:8099/company/" + action)
                    .params(params).accept(MediaType.ALL);
        }else{
            if(queryParams==null)queryParams=new LinkedMultiValueMap<>();
            request=MockMvcRequestBuilders.get("http://localhost:8080/company/"+action)
                    .queryParams(queryParams).accept(MediaType.ALL);
        }
        //.content(JSON.toJSONString(params));
        try {
            MvcResult mvcResult = this.mockMvc.perform(request)
                    .andDo(MockMvcResultHandlers.print())
                    //.andExpect(MockMvcResultMatchers.status().isOk())
                    .andReturn();
            int status=mvcResult.getResponse().getStatus();              //得到返回代码
            String content=mvcResult.getResponse().getContentAsString();  //得到返回结果
            Assert.assertEquals(200,status);                              //判断返回代码是否正确
            ObjectMapper objectMapper = new ObjectMapper();
            //校验返回信息
            log.info(content);
        }catch (Exception e){

        }
    }
    @Test
    void willAdd() {
        doAction("get","/show/1",null,null);
    }

    @Test
    void addCompany() {
        MultiValueMap params=new LinkedMultiValueMap<>();
        params.add("comName","dsd22");
        params.add("comType","dsd");
        params.add("comCreatetime","2015-02-12");
        params.add("comContactor","zhang");
        params.add("comAddr","北京");
        params.add("comEmail","dd@ssc.com");
        params.add("comTel","34343443");
        params.add("comFax","3332");
        params.add("comZip","343434");
        params.add("comEmpnum","50");
        params.add("comHomepage","http://dfdsdsf");
        params.add("comDesc","dssddsds");
        doAction("post","/add",params,null);
    }

    @Test
    void editCompany() {
        MultiValueMap params=new LinkedMultiValueMap<>();
```

```java
        params.add("comId","1");
        params.add("comName","dsd22");
        params.add("comType","dsd");
        params.add("comCreatetime","2015-02-12");
        params.add("comContactor","zhang");
        params.add("comAddr","北京");
        params.add("comEmail","dd@ssc.com");
        params.add("comTel","34343443");
        params.add("comFax","3332");
        params.add("comZip","343434");
        params.add("comEmpnum","50");
        params.add("comHomepage","http://dfdsdsf");
        params.add("comDesc","dssddsds");
        doAction("post","/edit",params,null);
    }

    @Test
    void delete() {
        doAction("get","/delete/2",null,null);
    }

    @Test
    void willEdit() {
        doAction("get","/willEdit/1",null,null);
    }

    @Test
    void show() {
        doAction("get","/show/1",null,null);
    }

    @Test
    void findCompanies() {
        MultiValueMap params=new LinkedMultiValueMap<>();
        params.add("comType","");
        params.add("comName","");
        params.add("pageNo","1");
        params.add("pageSize","10");
        doAction("post","/manage",params,null);
    }
}
```

2.6 视图层实现

视图层采用 JSP 实现，使用了标准标签库 JSTL。这里只给出系统首页和管理员视图设计，这些 JSP 文件均被存放在 WEB-INF/jsp 文件夹下。

2.6.1 系统首页设计

系统首页的界面效果如图 1-51 所示。界面的结构布局主要使用 DIV 标签并与 CSS 结合来实现，同时使用<c:choose>标签来判断用户是否登录，代码如下：

```
<%@ page contentType="text/html;charset=UTF-8" language="java" %>
<%@ taglib uri="http://java.sun.com/jsp/jstl/core" prefix="c" %>
<!DOCTYPE html>
<html>
<head>
    <meta http-equiv="Content-Type" content="text/html; charset=UTF-8">
    <title>网上人才中心系统</title>
    <link rel="stylesheet" href="/rc/static/css/all.css" type="text/css"/>
</head>
</head>
<body>
<div id="container">
    <div id="header"></div>
    <div id="menu">
        <ul>
            <li><a href="/rc " target="_top">【系统首页】</a></li>
            <li><a href="/rc/news/list" target="content">【站内新闻】</a></li>
            <li><a href="/rc/job/list" target="content">【职位信息】</a></li>
            <li><a href="/rc/company/list" target="content">【企业信息】</a></li>
            <li><a href="/rc/person/list" target="content">【人才信息】</a></li>
            <li><a href="/rc/person/center" target="content">【个人中心】</a></li>
        </ul>
    </div>
    <div id="pagebody">
        <div id="left" style="text-align: center">
            <br/>
            <c:choose>
                <c:when test="${session.user==null}">
                    <form name="form1" method="POST" action="/rc/user/login">
                        <p>用户名：<input type="text" name="userName" size="12" /></p>
                        <p>密 码：<input type="password" name="userPwd" size="12" /></p>
                        <p><a href="JavaScript:document.form1.submit()"/>登录</a>
                          <a href="JavaScript:document.location.href='/rc/user/willRegister'">注册</a></p>
                        <p><a href="/rc/user/findPwd">忘记密码</a></p>
                    </form>
                </c:when>
                <c:otherwise>
                    <br/>
                    <p>欢迎${user.userName}进入</p>
                    <p><a href="/rc/logout"/>注销</a></p>
                </c:otherwise>
            </c:choose>
```

```
            </div>
        <div id="right">
                <iframe id="content" name="content" frameborder="0" width="100%" height="100%" src="/rc/first1">
                </iframe>
            </div>
    </div>
    <div id="footer">
        <hr/>
        Copyright &copy; 2020-2030 YSL.All Rights Reserved.
    </div>
</div>
</body>
</html>
```

2.6.2 管理员视图设计

这里只创建系统管理首页和其中的企业管理视图。

1. 系统管理首页（jsp/manage.jsp）设计

当管理员登录后，即可访问系统管理首页，如图 2-12 所示。

图 2-12 系统管理首页

具体代码如下：

```
<%@ page contentType="text/html;charset=UTF-8" language="java" %>
<!DOCTYPE html>
<html>
<head>
    <meta http-equiv="Content-Type" content="text/html; charset=UTF-8">
```

```html
        <title>网上人才中心系统</title>
        <link rel="stylesheet" href="/rc/static/css/all.css" type="text/css"/>
</head>
</head>
<body>
<div id="container">
    <div id="header"></div>
    <div id="pagebody">
        <div id="left" style="text-align: center">
            <br/>
            <p><b>系统管理</b></p>
            <hr/>
            <p><a href="/rc/company/manage" target="content">企业管理</a></p>
            <p><a href="/rc/user/manage" target="content">用户管理</a></p>
            <p><a href="/rc/permission/manage" target="content">权限管理</a></p>
            <p><a href="/rc/role/manage" target="content">角色管理</a></p>
            <p><a href="/rc/role/manage" target="content">新闻管理</a></p>
            <p><a href="/rc/user/willEditPwd" target="content">修改密码</a></p>
            <p><a href="/rc/logout">退出管理</a></p>
        </div>
        <div id="right">
            <iframe id="content" name="content" frameborder="0" width="100%"
                height="100%" src="/rc/first"></iframe>
        </div>
    </div>
    <div id="footer">
        <hr />Copyright &copy; 2020-2030 YSL.All Rights Reserved.
    </div>
</div>
</body>
</html>
```

2．企业管理视图设计

企业管理视图主要涉及企业管理页、企业添加页、企业修改页、查看企业页，这些JSP文件均被存放在 WEB-INF/jsp/company 文件夹下。

1）企业管理页（company/manage.jsp）

以管理员身份登录，进入系统管理首页，然后选择左侧的【企业管理】选项，即可打开企业管理页，运行效果如图2-13所示。在企业管理页中，可以浏览、查询、添加、修改、删除已加入的企业信息。

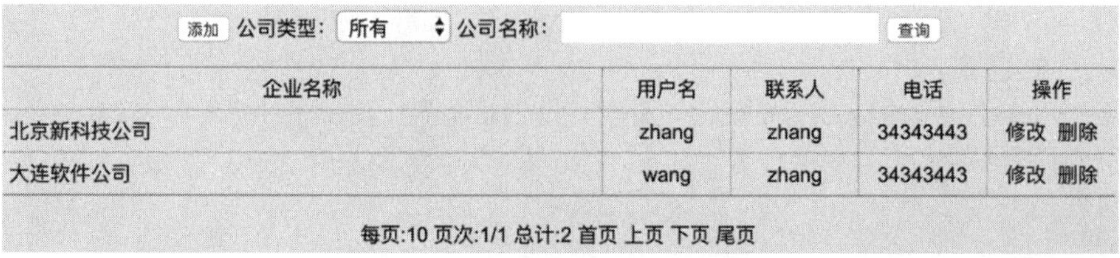

图2-13　企业管理页的运行效果

具体代码如下：
```jsp
<%@ page contentType="text/html;charset=UTF-8" language="java" %>
<%@ taglib uri="http://java.sun.com/jsp/jstl/core" prefix="c" %>
<!DOCTYPE html>
<html>
<head>
    <meta http-equiv="Content-Type" content="text/html; charset=UTF-8"/>
    <title>公司管理</title>
    <link rel="stylesheet" href="/rc/static/css/all.css" type="text/css"/>
</head>
<body>
<br/>
<form method="post" action="/rc/manage">
    <input type="button" value="添加" onclick="window.location.href='/rc/company/willAdd'"/>
    公司类型：<select name="comType">
    <option>所有</option>
    <option>国有企业</option>
    <option>民营企业</option>
    <option>合资企业</option>
    <option>外资企业</option>
    </select>
    公司名称：<input type="text" name="comName" size="30" value="${comName}"/>
    <input type="submit" value="查询"/>
</form>
<table class="mytable1" align="center" cellpadding="4" cellspacing="0">
    <tr>
        <th style="border-left-width:0px">企业名称</th>
        <th width="80">用户名</th>
        <th width="80">联系人</th>
        <th width="80">电话</th>
        <th width="80">操作</th>
    </tr>
    <c:forEach items="${map['list'] }" var="n">
        <tr>
            <td style="text-align:left;border-left-width:0px">
                <a href="/rc/show/${n.userId}">${n.comName} </a>
            </td>
            <td>${n.userName}</td>
            <td>${n.comContactor}</td>
            <td>${n.comTel} </td>
            <td align="center">
                <a href="/rc/company/willEdit/${n.userId}">修改</a> 
                <a href="/rc/company/delete/${n.userId}">删除</a>
            </td>
        </tr>
    </c:forEach>
</table>
<br/>
<center> ${map['pageHelper'] } </center>
```

</body>
</html>

2）企业添加页（company/add.jsp）

在企业管理页中单击【添加】按钮，即可进入企业添加页，运行效果如图2-14所示。

图2-14 企业添加页的运行效果

具体代码如下：

```jsp
<%@ page contentType="text/html;charset=UTF-8" language="java" %>
<!DOCTYPE html>
<html>
    <head>
        <meta http-equiv="Content-Type" content="text/html; charset=UTF-8">
        <title>添加企业</title>
        <link rel="stylesheet" href="/rc/static/css/all.css" type="text/css"/>
    </head>
    <body>
        <form method="POST" action="/rc/company/add">
            <table class="mytable2" align="center" cellspacing="0" cellpadding="2">
                <tr>
                    <th>用户名</th>
                    <td><input type="text" name="userName" size="18"></td>
                </tr>
                <tr>
                    <th>密码</th>
                    <td><input type="password" name="userPwd" size="18"></td>
                </tr>
                <tr>
```

```html
            <th>确认密码</th>
            <td><input type="password" name="userPwd1" size="18"></td>
        </tr>
        <tr>
            <th>公司名称</th>
            <td><input type="text" name="comName" size="50"></td>
        </tr>
        <tr>
            <th>企业类型</th>
            <td><select size="1" name="comType">
                    <option selected value="国有企业">国有企业</option>
                    <option value="民营企业">民营企业</option>
                    <option value="合资企业">合资企业</option>
                    <option value="外资企业">外资企业</option>
                </select></td>
        </tr>
        </tr>
        <tr>
            <th>创办时间</th>
            <td><input type="text" name="comCreatetime" size="20"></td>
        </tr>
        <tr>
            <th>联系人</th>
            <td><input type="text" name="comContactor" size="30"></td>
        </tr>
        <tr>
            <th>员工数</th>
            <td><input type="text" name="comEmpnum" size="5"></td>
        </tr>
        <tr>
            <th>电话</th>
            <td><input type="text" name="comTel" size="20"></td>
        </tr>
        <tr>
            <th>传真</th>
            <td><input type="text" name="comFax" size="20"></td>
        </tr>
        <tr>
            <th>地址</th>
            <td><input type="text" name="comAddr" size="70"></td>
        </tr>
        <tr>
            <th>邮编</th>
            <td><input type="text" name="comZip" size="10"></td>
        </tr>
        <tr>
            <th>公司主页</th>
            <td><input type="text" name="comHomepage" size="10"></td>
        </tr>
```

```
                <tr>
                    <th>Email</th>
                    <td><input type="text" name="comEmail" size="70"></td>
                </tr>
                <tr>
                    <th>简介</th>
                    <td><textarea rows="6" name="comDesc" cols="55"></textarea></td>
                </tr>
            </table>
            <p align="center">
                <input type="submit" value=" 提 交 ">
                <input type="button" value=" 取 消 " onclick="window.history.back();">
            </p>
        </form>
    </body>
</html>
```

3）企业修改页（company/edit.jsp）

在管理员登录后，进入企业管理页，并选择某行的【修改】选项，即可打开企业修改页，运行效果如图 2-15 所示。

图 2-15　企业修改页的运行效果

具体代码如下：

```
<%@ page contentType="text/html;charset=UTF-8" language="java" %>
<!DOCTYPE html>
<html>
```

```html
<head>
    <meta http-equiv="Content-Type" content="text/html; charset=UTF-8">
    <title>修改企业</title>
    <link rel="stylesheet" href="/rc/static/css/all.css" type="text/css"/>
</head>
<body>
<form method="POST" action="/rc/company/edit">
    <input type="hidden" name="userId" value="${company.userId}">
    <table class="mytable2" align="center" cellspacing="0" cellpadding="2">
        <tr>
            <th>用户名</th>
            <td>
                <input type="text" name="userName" size="18"
                                   value="${company.userName}"></td>
        </tr>
        <tr>
            <th>密码</th>
            <td><input type="password" name="userPwd" size="18"
                                   value="${company.userPwd}"></td>
        </tr>
        <tr>
            <th>确认密码</th>
            <td><input type="password" name="userPwd1" size="18"
                                   value="${company.userPwd}"></td>
        </tr>
        <tr>
            <th>公司名称</th>
            <td><input type="text" name="comName" size="50"
                                   value="${company.comName}"></td>
        </tr>
        <tr>
            <th>企业类型</th>
            <td>
                <select size="1" name="comType">
                    <option value="国有企业">国有企业</option>
                    <option value="民营企业">民营企业</option>
                    <option value="合资企业">合资企业</option>
                    <option value="外资企业">外资企业</option>
                </select>
                <script>
                    var sel=document.getElementsByName("comType")[0];
                    var s='${company.comType}';
                    for(var i=0;i<sel.options.length;i++){
                        if(sel.options[i].value==s){
                            sel.options[i].selected=true;
                            break;
                        }
                    }
                </script>
```

```html
                </td>
        </tr>
        </tr>
        <tr>
            <th>创办时间</th>
            <td><input type="text" name="comCreatetime" size="20"
                            value="${company.comCreatetime}"></td>
        </tr>
        <tr>
            <th>联系人</th>
            <td>
                <input type="text" name="comContactor" size="30"
                            value="${company.comContactor}"></td>
        </tr>
        <tr>
            <th>员工数</th>
            <td><input type="text" name="comEmpnum" size="5"
                            value="${company.comEmpnum}"></td>
        </tr>
        <tr>
            <th>电话</th>
            <td><input type="text" name="comTel" size="20" value="${company.comTel}">
            </td>
        </tr>
        <tr>
            <th>传真</th>
            <td><input type="text" name="comFax" size="20" value="${company.comFax}">
            </td>
        </tr>
        <tr>
            <th>地址</th>
            <td><input type="text" name="comAddr" size="70" value="${company.comAddr}">
            </td>
        </tr>
        <tr>
            <th>邮编</th>
            <td><input type="text" name="comZip" size="10" value="${company.comZip}">
            </td>
        </tr>
        <tr>
            <th>公司主页</th>
            <td><input type="text" name="comHomepage" size="10"
                        value="${company.comHomepage}"></td>
        </tr>
        <tr>
            <th>Email</th>
            <td><input type="text" name="comEmail" size="70"
                            value="${company.comEmail}"></td>
        </tr>
```

```html
            <tr>
                <th>简介</th>
                <td><textarea rows="6" name="comDesc"cols="55">
                                ${company.comDesc}</textarea></td>
            </tr>
        </table>
        <p align="center">
            <input type="submit" value=" 提 交 ">
            <input type="button" value=" 取 消 " onclick="window.history.back();">
        </p>
    </form>
</body>
</html>
```

4）查看企业页（company/show.jsp）

在管理员登录后，可以在企业管理页中单击企业名称，打开查看企业页来查看企业的详细信息。查看企业的运行效果如图 2-16 所示。

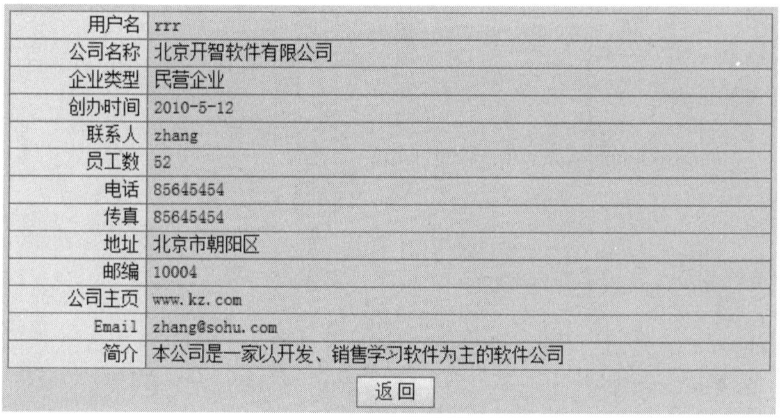

用户名	rrr
公司名称	北京开智软件有限公司
企业类型	民营企业
创办时间	2010-5-12
联系人	zhang
员工数	52
电话	85645454
传真	85645454
地址	北京市朝阳区
邮编	10004
公司主页	www.kz.com
Email	zhang@sohu.com
简介	本公司是一家以开发、销售学习软件为主的软件公司

返回

图 2-16 查看企业页的运行效果

具体代码如下：

```jsp
<%@ page contentType="text/html;charset=UTF-8" language="java" %>
<!DOCTYPE html>
<html>
    <head>
        <meta http-equiv="Content-Type" content="text/html; charset=UTF-8">
        <title>查看企业</title>
        <link rel="stylesheet" href="/rc/static/css/all.css" type="text/css"/>
    </head>
    <body>
        <table class="mytable2" align="center" cellspacing="0" cellpadding="2">
            <tr>
                <th>用户名</th>
                <td>
                    ${company.userName}</td>
            </tr>
```

```html
            <tr>
                <th>公司名称</th>
                <td>${company.comName}</td>
            </tr>
            <tr>
                <th>企业类型</th>
                <td>
                    ${company.comType}
                </td>
            </tr>
        </tr>
        <tr>
            <th>创办时间</th>
            <td>${company.comCreateTime}</td>
        </tr>
        <tr>
            <th>联系人</th>
            <td>
                ${company.comContactor}</td>
        </tr>
        <tr>
            <th>员工数</th>
            <td>${company.comEmpNum}</td>
        </tr>
        <tr>
            <th>电话</th>
            <td>${company.comTel}</td>
        </tr>
        <tr>
            <th>传真</th>
            <td>${company.comFax}</td>
        </tr>
        <tr>
            <th>地址</th>
            <td>${company.comAddr}</td>
        </tr>
        <tr>
            <th>邮编</th>
            <td>${company.comZip}</td>
        </tr>
        <tr>
            <th>公司主页</th>
            <td>${company.comHomepage}</td>
        </tr>
        <tr>
            <th>Email</th>
            <td>${company.comEmail}</td>
        </tr>
        <tr>
```

```
            <th>简介</th>
            <td>${company.comDesc}</td>
        </tr>
    </table>
    <p align="center">
        <input type="button" value=" 返 回 " onclick="window.history.back();">
    </p>
</body>
</html>
```

2.6.3 部署运行程序

1．使用内置的 Tomcat 运行

单击如图 2-17 所示的工具栏中的运行按钮，即可运行程序。

图 2-17　工具栏

或者通过浏览器访问 http://localhost:8080/rc。

2．使用外置的 Tomcat 运行

默认运行列表中只有主类，如果想使用外置的 Tomcat 运行，则可按以下步骤进行配置。

（1）单击运行列表旁的下拉按钮，然后从下拉列表中选择【Edit Configures】选项，打开运行配置界面，如图 2-18 所示。

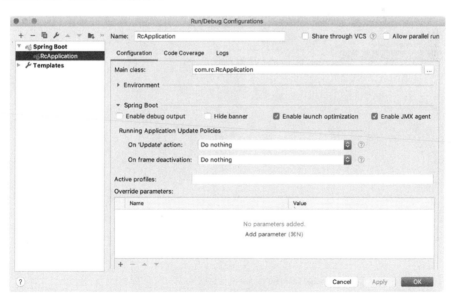

图 2-18　运行配置界面

(2)在运行配置界面中单击左上角的【+】按钮,再从左侧的列表框中选择【Tomcat Server】→【Unnamed】选项,设置服务器,如图 2-19 所示。然后选择【Deployment】选项卡。

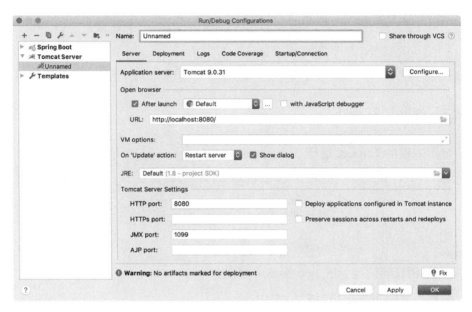

图 2-19　设置服务器

(3)在如图 2-20 所示的界面中设置发布程序。单击下面的【+】按钮,在弹出的列表中选择【Artifact】选项。

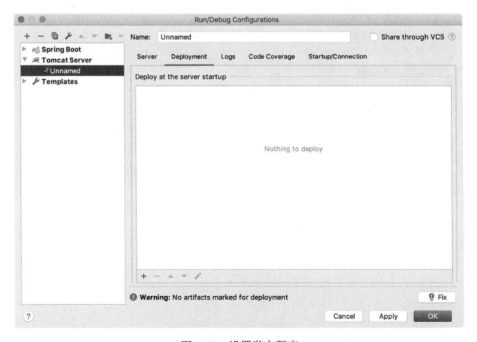

图 2-20　设置发布程序

（4）在如图 2-21 所示的界面中选择要发布的 war 包，单击【OK】按钮。返回到运行配置界面，单击【OK】按钮。

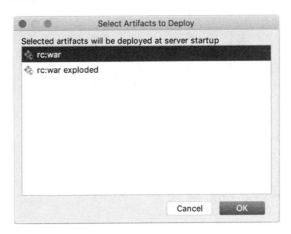

图 2-21　选择要发布的 war 包

（5）在运行列表中会出现所配置的选项，选择该选项并单击运行按钮，即可运行程序。

第 3 章 基于 Spring Data JPA 实现数据访问层

本章主要介绍 Spring Data JPA 相关技术，包括 Spring Data JPA 概述、Spring Boot 与 Spring Data JPA 整合、实体对象映射和 JPA 数据操作方法。

3.1 Spring Data JPA 概述

3.1.1 ORM 与 JPA

1. 持久化技术与 ORM

简单来说，持久化就是将内存中的数据保存到关系数据库、文件系统、消息队列等提供持久化支持的设备中。数据访问层就是系统中专注于实现数据持久化的相对独立的层。在分层结构设计中，数据访问层位于数据库层与业务逻辑层之间，实现了业务逻辑与数据库层的分离，给开发人员提供了便利。

面向对象的开发方法是当今企业级应用开发环境中的主流开发方法，关系数据库是企业级应用环境中永久存放数据的主流数据存储系统。对象和关系数据是业务实体的两种表现形式：业务实体在内存中表现为对象，在数据库中表现为关系数据。对象关系映射（Object Relational Mapping，简称 ORM）通过使用描述对象和数据库之间映射的元数据，可以将面向对象语言程序中的对象自动持久化到关系数据库中，其本质上就是将数据从一种形式转换到另一种形式。因此，ORM 框架是解决面向对象的程序设计语言与关系数据库间存在不匹配问题的中间方案。Hibernate、TopLink 及 OpenJP 都属于 ORM 框架。ORM 框架的出现，使得开发者从数据库编程中解脱出来，把更多的精力放在业务模型与业务逻辑上。

2. JPA

在 JPA（Java Persistence API，Java 持久化接口）规范出现之前，由于没有官方的标准，因此各 ORM 框架之间的 API 差别很大，使用了某种 ORM 框架的系统会严重受制于该 ORM 的标准。

JPA 是 Sun 公司提出的持久化解决规范，它充分吸收了 Hibernate、TopLink、JDO 等 ORM 框架的优势，具有易于使用、伸缩性强的特点。JPA 设计的主要目标是简化现有的持久化开发工作，并为 ORM 技术提供一套标准化的接口。JPA 通过标注或 XML 描述对象关系的映射，并将运行期的实体对象持久化到数据库中，其结构如图 3-1 所示。

第 3 章 基于 Spring Data JPA 实现数据访问层

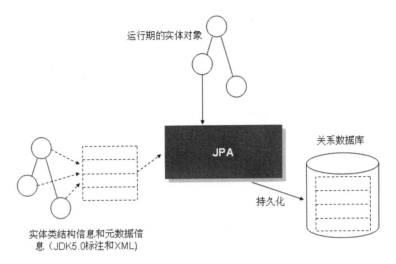

图 3-1　JPA 结构

JPA 包括以下 3 个方面的内容。

- 一套 API 标准。该标准被存放在 javax.persistence 的包下面，用来操作实体对象，执行 CRUD 操作。框架会在后台代替我们完成所有的事情，使开发者从烦琐的 JDBC 和 SQL 代码中解脱出来。
- JPQL（Java Persistence Query Language，面向对象的查询语言）。这是持久化操作中一个很重要的方面，通过面向对象而非面向数据库的查询语言查询数据，可以避免程序中的 SQL 语句紧密耦合。
- ORM。JPA 支持 XML 和 JDK5.0 标注两种元数据的形式。元数据用于描述对象和表之间的映射关系，框架可以据此将实体对象持久化到数据库表中。

JPA 从以下几个方面显示出其强大的优势：标准化，对容器级特性的支持，简单易用，集成方便，强大的查询能力，支持面向对象的高级特性。

JPA 本质上是一种 ORM 规范，但并未提供 ORM 实现，其具体实现由其他厂商来提供。程序员若使用 JPA，则仍然需要选择 JPA 的实现框架。就目前来说，对 JPA 规范实现最好的就是 Hibernate。Hibernate 是一个面向 Java 环境的对象关系数据库映射工具，即 ORM 工具。它对 JDBC API 进行了封装，不仅可以管理 Java 类到数据库表的映射，还提供了查询和获取数据的方法，可以大幅度减少开发时人工使用 SQL 和 JDBC 处理数据的时间。从 Hibernate 3.2 以后的版本就开始支持 JPA。

3.1.2　Spring Data JPA

1．Spring Data 项目

Spring Data 是从 2010 年发展起来的项目，从创立之初，Spring Data 就想给大家提供一个熟悉的、一致的、基于 Spring 的数据访问编程模型，同时仍然保留底层数据存储的特性。它可以轻松地让开发者使用数据访问技术，包括关系数据库、非关系数据库（NoSQL）和基于云的数据服务。

Spring Data Common 是 Spring Data 所有模块的公用部分，用于提供跨 Spring 数据项目的共享基础设施。它包含了技术中立的库接口及一个支持 Java 类的元数据模型。

Spring Data 不仅对传统的数据库访问技术 JDBC、Hibernate、JDO、TopLink、JPA、MyBatis 提供了很好的支持、扩展、抽象和方便的 API，还对 NoSQL 等非关系数据提供了很好的支持，包括 MongoDB、Redis、Apache Solr 等。

2. Spring Data JPA 项目

Spring Data JPA 是 Spring Data 一个子项目，目的是简化 JPA 开发。我们可以将它理解为 JPA 规范的再次封装抽象，其底层仍然使用了 Hibernate 的 JPA 技术实现，与 JPA 的关系如图 3-2 所示。在基于这个框架编写数据访问层时，只需定义 Dao 接口继承于它内部定义的接口，就可以在不编写接口实现类的情况下，实现对数据库的访问和操作，同时可以提供很多除 CRUD 操作外的功能，如分页、排序、复杂查询等。Spring Data JPA 让我们从 Dao 层的操作中解脱出来，几乎所有的 CRUD 操作都可以依赖于它来实现。

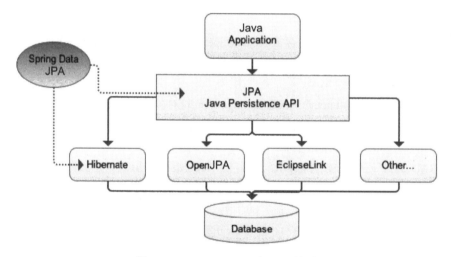

图 3-2　Spring Data JPA 与 JPA 关系

随着 Spring Boot 和 Spring Cloud 在市场上的流行，Spring Data JPA 也逐渐进入大家的视野，将它们组合成有机的整体，使用起来会比较方便，不但可以加快开发的效率，而且可以使开发者不需要关心和配置更多的东西，完全沉浸在 Spring 的完整生态标准下。由于 JPA 上手简单，开发效率高，对对象的支持比较好，并且具有很大的灵活性，因此市场对其的认可度越来越高。

3.1.3　Spring Data JPA 接口和类

Spring Data JPA 提供了一整套接口和类。使用这些接口和类可以很容易地实现 Dao，这些接口和类的层次结构如图 3-3 所示。

第3章 基于 Spring Data JPA 实现数据访问层

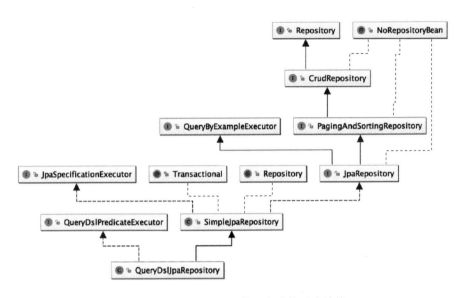

图 3-3 Spring Data JPA 接口和类的层次结构

在下面的介绍中,范型参数 T 是要操作的实体类,ID 是实体类主键的类型。

(1) Repository:顶层的接口,是一个空接口,目的是统一所有的 Repository 的类型,并且能够在扫描组件时自动识别。

(2) CrudRepository:继承于 Repository 接口,提供了 11 个 CRUD 操作的常用方法。

- <S extends T> S save(S entity);//保存
- <S extends T> Iterable<S> save(Iterable<S> entities);//批量保存
- T findOne(ID id);//根据 ID 查询一个对象。返回对象本身,当对象不存在时,返回 null
- Iterable<T> findAll();//查询所有对象
- Iterable<T> findAll(Iterable<ID> ids);//根据 ID 列表查询所有对象
- boolean exists(ID id);//根据 ID 判断对象是否存在
- long count();//计算对象的总数
- void delete(ID id);//根据 ID 删除对象
- void delete(T entity);//删除一个对象
- void delete(Iterable<? extends T> entities);//批量删除集合对象(在后台执行时,一条一条地删除)
- void deleteAll();//删除所有对象(在后台执行时,一条一条地删除)

(3) PagingAndSortingRepository:继承于 CrudRepository 接口,增加了两个方法,实现了分页和排序的功能。

- Iterable<T> findAll(Sort sort);//查询所有对象,仅排序
- Page<T>findAll(Pageable pageable);// 查询所有对象,分页和排序

(4) QueryByExampleExecutor:支持根据实例查询对象。

- <S extends T> Optional<S> findOne(Example<S> var1);//根据实例查询一个对象
- <S extends T> Iterable<S> findAll(Example<S> var1);//根据实例查询所有对象

- `<S extends T> Iterable<S> findAll(Example<S> var1, Sort var2);` //根据实例查询所有对象并排序
- `<S extends T> Page<S> findAll(Example<S> var1, Pageable var2);` //根据实例查询所有对象，分页并排序
- `<S extends T> long count(Example<S> var1);` //根据实例查询对象总数
- `<S extends T> boolean exists(Example<S> var1);` //查询实例是否存在

（5）JpaRepository：继承于 PagingAndSortingRepository 和 QueryByExampleExecutor 接口，并增加了几个查询及批量保存方法、重写了 QueryByExampleExecutor 接口的两个方法。查询所返回的是 List，使用起来更方便。另外，增加了 InBatch 删除，在实际执行时，后台会生成一条 SQL 语句，效率相比 CrudRepository 接口更高；增加了 getOne() 方法，该方法返回的是对象引用，当查询的对象不存在时，它的值不是 null。

- `List<T> findAll();` //查询所有对象，返回 List
- `List<T> findAll(Sort sort);` //查询所有对象，并排序，返回 List
- `List<T> findAll(Iterable<ID> ids);` //根据 ID 列表查询所有对象，返回 List
- `void flush();` //强制缓存与数据库同步
- `<S extends T> List<S> save(Iterable<S> entities);` //批量保存，并返回 List
- `<S extends T> S saveAndFlush(S entity);` //保存并强制同步数据库
- `void deleteInBatch(Iterable<T> entities);` //批量删除集合对象（在后台执行时，生成一条语句，使用多个 or 条件）
- `void deleteAllInBatch();` //删除所有对象（执行一条语句）
- `T getOne(ID id);` //根据 ID 查询一个对象，返回对象的引用（区别于 findOne）。当对象不存在时，返回的引用不是 null，但各个属性值是 null
- `<S extends T> List<S> findAll(Example<S> example);` //根据实例查询对象
- `<S extends T> List<S> findAll(Example<S> example, Sort sort);` //根据实例查询对象，并排序

（6）JpaSpecificationExecutor：复杂查询的接口。可通过 Specification 参数实现复杂查询，支持分页和排序。

- `Optional<T> findOne(@Nullable Specification<T> var1);` //根据动态条件查询一个对象，返回一个 Optional 对象
- `List<T> findAll(@Nullable Specification<T> var1);` //根据动态条件查询所有对象，返回 List
- `Page<T> findAll(@Nullable Specification<T> var1, Pageable var2);` //根据动态条件查询所有对象，分页并排序，返回 List
- `List<T> findAll(@Nullable Specification<T> var1, Sort var2);` //根据动态条件查询所有对象并排序，返回 List
- `long count(@Nullable Specification<T> var1);` //根据动态条件查询对象总数

（7）JpaRepositoryImplementation：继承于 JpaRepository、JpaSpecificationExecutor 接口，增加了设置持久化方法元数据和设置 ESCAPE 字符的方法。

（8）QueryDslPredicateExecutor：用于支持 QueryDSL 的查询操作接口。QueryDSL 定义了一种常用的静态类型语法，用于在持久域模型数据之上进行查询。它是一个通用的查询框架，专注于通过 Java API 构建安全的 SQL 查询。

（9）SimpleJpaRepository：JpaRepositoryImplementation 接口实现类。

（10）QueryDslJpaRepository：继承于 SimpleJpaRepository 类，并实现了 QueryDslPredicateExecutor 接口。

3.2 Spring Boot 与 Spring Data JPA 整合

3.2.1 Spring Data JPA 基本配置

1. 引入依赖

使用 Spring Data JPA 需先引入如下依赖：

```xml
<!-- JPA 集成 -->
<dependency>
    <groupId>org.springframework.boot</groupId>
    <artifactId>spring-boot-starter-data-jpa</artifactId>
</dependency>
```

2. 配置 JPA

```
#数据源配置
spring.datasource.url=jdbc:mysql://localhost:3306/rc?useUnicode=true&characterEncoding=utf8&serverTimezone=GMT&useSSL=false&zeroDateTimeBehavior=convertToNull&allowMultiQueries=true
spring.datasource.username=root
spring.datasource.password=12345
spring.datasource.driverClassName=com.mysql.cj.jdbc.Driver
spring.datasource.type=com.alibaba.druid.pool.DruidDataSource
#JPA 相关配置
spring.jpa.database = MYSQL
spring.jpa.database-platform=org.hibernate.dialect.MySQL5InnoDBDialect
spring.jpa.show-sql=true
spring.jpa.hibernate.naming-strategy = org.hibernate.cfg.ImprovedNamingStrategy
spring.jpa.hibernate.ddl-auto=update
#解决延迟加载问题
spring.jpa.open-in-view=true
spring.jpa.properties.hibernate.enable_lazy_load_no_trans=true
```

在上述配置中，spring.jpa.hibernate.ddl-auto 可取的值有以下几种。

- create：在每次加载 Hibernate 时，都会删除上一次生成的表，然后根据 Model 类重新生成新表。即使两次没有任何改变，也要这样执行，这就是导致数据库表数据丢失的一个重要原因。
- create-drop：在每次加载 Hibernate 时，都会根据 Model 类生成新表，但是 sessionFactory 一关闭，就会自动删除表。

- update：最常用的属性，在第一次加载 Hibernate 时，会根据 Model 类自动建立表结构（前提是先建立好数据库），在以后加载 Hibernate 时，会根据 Model 类自动更新表结构。虽然表结构改变了，但表中的行仍然存在，不会删除以前的行。需要注意的是，当部署到服务器后，表结构是不会被马上建立起来的，需要等到应用第一次运行起来后才可以建立。
- validate：在每次加载 Hibernate 时，都会验证创建的数据库表结构。只会和数据库中的表进行比较，不会创建新表，但是会插入新值。

3.2.2 数据源配置优化

1. 连接池技术概述

数据库连接是一种关键的、有限的、昂贵的资源，这在多用户的网页应用程序中体现得尤为明显。一个数据库连接对象对应一个物理数据库连接，在每次操作时都打开一个物理数据库连接，并在使用完成后关闭，这样会造成系统的性能较低。数据库连接池的解决方案是在应用程序启动时建立足够的数据库连接，并将这些连接组成一个连接池（简单来说，就是在一个"池"里存放很多半成品的数据库连接对象），由应用程序动态地对连接池中的连接进行申请、使用和释放。而多于连接池中连接数的并发请求应该在请求队列中排队等待。另外，应用程序可以根据连接池中连接的使用率，动态增加或减少连接池中的连接数。连接池技术尽可能多地重用了消耗内存的资源，大大节省了内存，提高了服务器的服务效率，能够支持更多的客户服务。使用连接池将大大提高程序运行效率，同时，可以通过其自身的管理机制来监视数据库连接的数量、使用情况等。

连接池技术可以大大地优化程序性能。然而，不同的连接池在并发请求压力下的稳定性等各有不同。在前些年，比较热门的连接池项目有 BoneCP、DBCP、C3P0 等。其中，BoneCP 在前几年非常流行，不过现在开始渐渐"没落"，就连 BoneCP 项目的作者也推荐用户使用 HikariCP。

HikariCP 是一个开源的数据库连接池组件，具有轻量级的代码，并且速度非常快。HikariCP 官网的性能测试对比如图 3-4 所示。

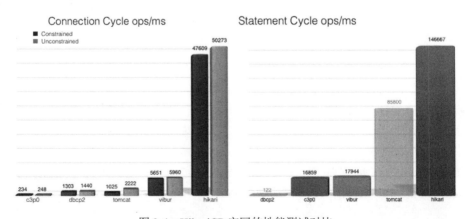

图 3-4　HikariCP 官网的性能测试对比

HikariCP 有以下几个特点。
- 字节码更加精简，可以加载更多代码到缓存中。
- 实现了一个无锁的集合类型，可以减少并发造成的资源竞争。
- 使用了自定义的数组类型，相对于 ArrayList 来说极大地提升了性能。
- 针对 CPU 的时间片算法进行了优化，尽可能在一个时间片中完成各种操作。

另外，HikariCP 拥有强劲的性能和稳定性，再加上其轻量级的代码，在当前"云时代、微服务"的背景下，得到了越来越多的人青睐。

Druid 是阿里巴巴开源平台上的一个数据库连接池实现，它结合了 C3P0、DBCP、PROXOOL 等数据库连接池的优点，同时加入了日志监控，可以很好地监控数据库连接池和 SQL 的执行情况，可以说是针对监控而生的数据库连接池。

Druid 支持所有 JDBC 兼容的数据库，包括 Oracle、MySQL、Derby、PostgreSQL、SQLServer、H2 等。Druid 针对 Oracle 和 MySQL 进行了特别优化，比如，Oracle 的 PSCache 内存占用优化，MySQL 的 ping 检测优化。Druid 在监控、可扩展性、稳定性和性能方面都有明显的优势。Druid 提供了 Filter-Chain 模式的扩展 API，可以自己编写 Filter 拦截 JDBC 中的任何方法，也可以在上面进行任何操作，如性能监控、SQL 审计、用户名和密码加密、日志查看等。

相比而言，HikariCP 更为轻量，性能比 Druid 高，而且 HikariCP 由 Spring Boot 2+官方支持，和 Spring Boot 兼容性更好。Druid 的优势是监控完善，扩展性更好。在使用时，我们可以根据需要选择，这里给出两种连接池的配置方法。

2. 使用 HikariCP

从 Spring Boot 2.0 开始使用 HikariCP，将默认的数据库连接池从 Tomcat Jdbc Pool 改为了 HikariCP。如果用户使用 spring-boot-starter-jdbc 或 spring-boot-starter-data-jpa，则会自动添加对 HikariCP 的依赖，也就是说，此时使用 HikariCP，我们只需配置连接池的其他参数即可。例如：

```
#数据源配置
spring.datasource.url=jdbc:mysql://localhost:3306/rc?useUnicode=true&characterEncoding=utf8&serverTimezone=GMT&useSSL=false&zeroDateTimeBehavior=convertToNull&allowMultiQueries=true
spring.datasource.username=root
spring.datasource.password=12345
spring.datasource.driverClassName=com.mysql.cj.jdbc.Driver
#客户端等待连接池连接的最大毫秒数
spring.datasource.hikari.connection-timeout=20000
#HikariCP 在连接池中维护的最小空闲连接数
spring.datasource.hikari.minimum-idle=5
#配置最大连接池大小
spring.datasource.hikari.maximum-pool-size=12
#允许连接在连接池中空闲的最长时间（以毫秒为单位）
spring.datasource.hikari.idle-timeout=300000
#连接池中连接关闭后的最长生命周期（以毫秒为单位）
spring.datasource.hikari.max-lifetime=1200000
#配置从连接池返回的连接的默认自动提交行为
```

spring.datasource.hikari.auto-commit=true

3．使用 Druid

使用 Druid 需先引入如下依赖：

```xml
<dependency>
    <groupId>com.alibaba</groupId>
    <artifactId>druid</artifactId>
    <version>1.1.10</version>
</dependency>
<dependency>
    <groupId>com.alibaba</groupId>
    <artifactId>druid-spring-boot-starter</artifactId>
    <version>1.1.10</version>
</dependency>
```

Druid 主要配置如下：

```
#数据源配置
spring.datasource.url=jdbc:mysql://localhost:3306/rc?useUnicode=true&characterEncoding=utf8&serverTimezone=GMT&useSSL=false&zeroDateTimeBehavior=convertToNull&allowMultiQueries=true
    spring.datasource.username=root
    spring.datasource.password=12345
    spring.datasource.driverClassName=com.mysql.cj.jdbc.Driver
    spring.datasource.type=com.alibaba.druid.pool.DruidDataSource
#连接池的配置信息
#在初始化时建立物理连接的个数
    spring.datasource.druid.initial-size=3
#连接池最小连接数
    spring.datasource.druid.min-idle=3
#连接池最大连接数
    spring.datasource.druid.max-active=20
#获取连接时的最大等待时间，单位为毫秒
    spring.datasource.druid.max-wait=60000
#申请连接的时候检测，如果空闲时间大于 timeBetweenEvictionRunsMillis，则执行 validationQuery
#检测连接是否有效
    spring.datasource.druid.test-while-idle=true
#既作为检测的间隔时间又作为 testWhileIdel 执行的依据
    spring.datasource.druid.time-between-connect-error-millis=60000
#在销毁线程时检测当前连接的最后活动时间和当前时间差大于该值时，关闭当前连接
    spring.datasource.druid.min-evictable-idle-time-millis: 30000
#用来检测连接是否有效的 SQL 语句必须是一个查询语句
#MySQL 中为 select 'x'
#Oracle 中为 select 1 from dual
    spring.datasource.druid.validation-query=select 'x'
#在申请连接时会执行 validationQuery，检测连接是否有效，若开启则会降低性能，默认为 true
    spring.datasource.druid.test-on-borrow=false
#在归还连接时会执行 validationQuery，检测连接是否有效，若开启则会降低性能，默认为 true
    spring.datasource.druid.test-on-return=false
#是否缓存 preparedStatement，MySQL 5.5+建议开启
    spring.datasource.druid.pool-prepared-statements=true
#当值大于 0 时，poolPreparedStatements 会自动修改为 true
```

spring.datasource.druid.max-pool-prepared-statement-per-connection-size=20
#合并多个 DruidDataSource 的监控数据
spring.datasource.druid.use-global-data-source-stat=false
#配置扩展插件
spring.datasource.druid.filters=stat,wall,slf4j
#通过 connect-properties 属性打开 mergeSql 功能，进行慢 SQL 记录
spring.datasource.druid.connect-properties = druid.stat.mergeSql=true;druid.stat.slowSqlMillis=5000
#定时输出统计信息到日志中，并且每次输出日志会导致连接池相关的计数器清零（reset）
spring.datasource.druid.time-between-log-stats-millis=300000
#配置 DruidStatFilter
spring.datasource.druid.web-stat-filter.enabled=true

上述配置使用了 Druid 监控功能，可以通过浏览器访问网址 http://localhost:8080/rc/druid/index.html，打开 Druid 监控页面，如图 3-5 所示。

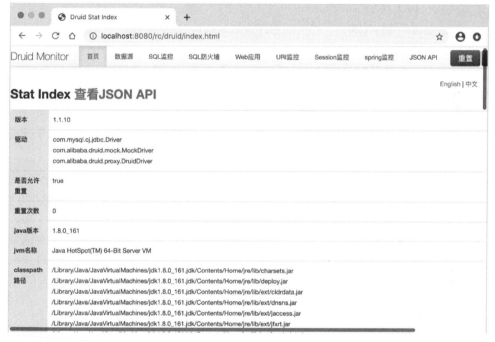

图 3-5　Druid 监控界面

3.2.3　基于 Spring Data JPA 实现 Dao 层

在 Spring Data JPA 中，通常使用 Repository 来进行数据层操作（作用相当于 Dao 层）。从图 3-3 中可以看出，JpaRepository 的父接口是 PagingAndSortingRepository 和 QueryByExampleExecutor；PagingAndSortingRepository 的父接口是 CrudRepository；CrudRepository 的父接口是 Repository。Repository 接口没有定义任何功能，它的作用就是标识。其中，JpaRepository 接口是最主要的接口，它不仅具有基本的 CRUD 功能，而且能够分页和排序，还可以按照实例进行查询。

在基于 Spring Data JPA 实现 Dao 层时，我们需要自定义接口继承于 JpaRepository 接

口，以实现所有的数据层基本操作功能。如果想实现复杂的查询操作，则还需要继承 JpaSpecificationExecutor 接口，这样可以建立一个规则来进行查询。例如：

> @Repository
> public interface RcAdminDao extends JpaRepository<RcAdmin,Integer>, JpaSpecificationExecutor<RcAdmin> {
>   RcAdmin findByUserNameAndUserPwd(String userName, String userPwd);
> }

  在有了接口之后，就不需要定义实现类了。这是因为，自定义的接口继承了 JpaRepository 接口，就相当于继承了 Repository 接口，那么 Spring Data JPA 在扫描时只要扫描到某个接口实现了 Repository 接口，就会自动地为其创建代理实现子类（通过 AOP 实现）。这个自动创建的实现子类 repository 就继承了 SimpleJpaRepository 类，从而可以完成数据库操作。

  其完成的过程大致如下所述。

  （1）引入依赖并进行配置后，在应用程序启动过程中，Spring Boot 的自动配置机制中关于 JPA 自动配置的部分就开始工作了，具体的自动配置类是 JpaRepositoriesAutoConfiguration。这里使用了 @EnableJpaRepositories 标注。@EnableJpaRepositories 标注指出了创建实现子类 repository 的 FactoryBean 是什么。默认使用 JpaRepositoryFactoryBean 为 JpaRepository 的 FactoryBean。JpaRepositoryFactory 类定义了如何生成最终 JpaRepository 的 Bean，并在生成时使用了实现类 SimpleJpaRepository。JpaRepositoryFactory 类继承于 RepositoryFactorySupport 类，其 getRepository() 方法也在该基类中实现。

  （2）getRepository() 方法创建了一个 SimpleJpaRepository 对象，该对象知道自己要针对哪个领域实体/数据库表进行操作，然后 getRepository() 方法又创建了该 SimpleJpaRepository 对象的一个代理对象，并附上有关的 Interceptor，返回该代理对象。最后，当真正的 JpaRepository 对象（也就是上面分析中所提到的 SimpleJpaRepository 代理对象）被创建之后，包裹该对象的那个 JpaRepositoryFactoryBean 对象就是最终我们要使用的 Bean 的 FactoryBean 了，并且在 Spring 容器中，对应用户自定义的 JpaRepository 接口，保存的 Bean 实际上是一个 JpaRepositoryFactoryBean。

3.2.4　Spring Data JPA 扩展

  在一般情况下，我们自定义接口继承于 JpaRepository 接口，就可以实现数据层基本操作功能。Spring Data JPA 具有按名称解析方法的特性，允许在自定义的接口中增加按特定命名规范书写的方法。此外，还可以使用 @Query 编写 JPQL 或原生 SQL。

  例如，使用 3 种方式实现同样的功能，代码如下：

> @Repository
> public interface RcAdminDao extends JpaRepository<RcUser,Integer>, JpaSpecificationExecutor<RcUser> {
>   //按命名
>   RcUser findByUserNameAndUserPwd(String userName, String userPwd);
>   //编写 JPQL
>   @Query("select a from RcUser a where a.userName=?1 and a.userPwd=?2")

```
        RcUser findByUserNameAndUserPwdJPQL(String userName, String userPwd);
        //编写原生 SQL
        @Query("select * from rc_user a where user_name=?1 and user_pwd=?2",nativeQuery = true)
        RcUser findByUserNameAndUserPwdNativeSQL(String userName, String userPwd);
}
```

上述类继承了 JpaSpecificationExecutor，因此对于比较复杂的查询，可在业务层以动态拼接查询逻辑形成 Specification 对象，并利用它调用 Dao 层。

虽然 Spring Data JPA 提供的功能已经足够强大了，但是依然不能满足实际开发中的所有需求。比如，不能处理超复杂的查询，而且比较复杂的查询代码在编写时也不是很"优雅"。各种复杂拼接查询逻辑都要编写在 Service 层，如果单独封装一个 Dao 类来编写复杂的查询，又显得有点多余和"臃肿"。因此，在一些特殊情况下，需要对 Spring Data JPA 中的方法进行重新实现或扩展。此外，有时可以创建一个自己的基接口，增加公用方法，同时继承 JpaRepository 和 JpaSpecificationExecutor，使其他接口继承这个基接口，从而简化代码书写，等等。

为了优雅并简化代码，在网上人才中心系统中，我们需要定义一个基接口，使其他接口继承该接口，并在基接口中增加一个或多个业务使用的公共方法。可以按以下步骤进行设计。

（1）声明一个接口，在该接口中声明需要自定义的方法，以及该接口需要继承的 Repository 接口或其子接口。比如，在网上人才中心系统中定义一个基接口，并在基接口中增加一个方法，代码如下：

```
@NoRepositoryBean
public interface RcBaseDao<T, ID extends Serializable> extends JpaRepositoryImplementation<T, ID> {
        List<T> findByNum(Integer num);
}
```

注意：全局扩展接口添加了@NoRepositoryBean 标注，告知 Spring Data 该实现类不是一个 Repository。

（2）定义一个类，用于实现上面的接口，该类继承于 SimpleJpaRepository 类，代码如下：

```
public class RcBaseDaoImpl<T, ID extends Serializable> extends SimpleJpaRepository<T,ID> implements RcBaseDao<T,ID>{
        private EntityManager entityManager;
        public BaseDaoImpl(Class<domainClass, EntityManager em) {
                super(domainClass, em);
                this.entityManager = em;
        }
}
@Override
public List<T> selectByNum(Integer num) {
        Sort sort = Sort.by(Sort.Direction.DESC, "comPosttime");
        Pageable pageable = PageRequest.of(0, num, sort);
        super.findAll(pageable).toList();
}
}
```

3.3 实体对象映射

3.3.1 实体映射基础

1．实体（Entity）

JPA 能够将普通的 Java 对象（有时被称为 POJO）映射到数据库，这些 Java 对象被称为 Entity（实体）。一个实体的定义符合 JavaBean 的规范，因此常常被称为实体 Bean。实体 Bean 的每个属性都被定义为私有的，并且有对应的 setter 和 getter 方法。JPA 将实体映射到数据库有两种方法：一种是基于 XML 文件；另一种是基于标注。目前，采用基于标注的方法映射实体更为普遍。

采用基于标注的方法映射的实体 Bean，在类的定义中还包含了映射信息，这些映射信息都是通过标注表示的。一个普通的 Java 类通过@Entity 标注可以映射为可持久化的实体，实体对应于数据库中的数据表。例如，职位的实体 Bean：

```
@Entity
@Table(name = "rc_job" , schema = "rc")
public class RcJob {
    ...
}
```

@Table 标注用来表示所映射的表，可以标注在类前。该标注主要有以下属性。

- name 表示实体所对应的表的名称，默认表名为实体的名称。
- schema 和 catalog 表示数据库名和目录名，会根据不同的数据库类型有所不同。

2．属性映射

@Column 标注用来表示属性所映射的字段，可以标注在 getter 方法或属性前，其属性包括：

- name 表示字段名。
- unique 表示该字段是否为唯一标识，默认为 false。
- nullable 表示该字段的值是否可以为 null，默认为 true。
- insertable 表示在使用 insert 脚本插入数据时，是否需要插入该字段的值。
- updatable 表示在使用 update 脚本更新数据时，是否需要更新该字段的值。
- columnDefinition 表示在创建表时，该字段创建的 SQL 语句一般在通过@Entity 标注生成表定义时使用。
- table 表示当映射多个表时，指定表中的字段。默认值为主表的表名。
- length 表示字段的长度，当字段的类型为 varchar 时，该属性才有效，默认为 255 个字符。
- precision 和 scale 表示精度，当字段类型为 double 时，precision 表示数值的总长度，scale 表示小数点所占的位数。

例如：

```
@Basic(optional = false)        //不允许为空
@Column(name = "job_title")     //设置属性 catName 对应的字段为 cat_name
```

```
private String jobTitle;
```
也可以进行如下配置：
```
@Column(name = " job_title ", nullable = false, length=50)//设置属性 name 对应的字段为 name，长度为 50，非空
private String jobTitle;
```

3. 主键映射

主键（Primary Key）是实体的唯一表示，每个实体应至少有一个唯一标识的主键。@Id 标注用于表示主键。例如：
```
@Id
private Integer catId;
```
一旦标注了主键，该实体属性的值就可以被指定，也可以根据一些特定的规则自动生成。这涉及另一个@GeneratedValue 标注的使用。

@GeneratedValue 标注和@ID 标注配合使用，用来设定主键的生成策略。其属性 strategy 用于指定生成策略，在默认情况下，JPA 会自动选择一个最适合底层数据库的主键生成策略；属性 generator 用于指定将被生成器引用的策略名。例如：
```
@Column(name = "job_id")
@Id
@GeneratedValue(strategy = GenerationType.IDENTITY) //主键自增，注意，这种方式依赖于具体的数据库，如果数据库不支持自增主键，则这个类型是无法使用的
@Basic(optional = false)
private int jobId;
```
在 javax.persistence.GenerationType 枚举中定义了以下几种可供选择的策略。

- IDENTITY 表示自增键字段，Oracle 不支持这种方式。
- AUTO JPA 表示自动选择合适的策略，是默认选项。
- SEQUENCE 表示通过序列产生主键，还可以进一步通过@SequenceGenerator 标注来指定更详细的生产方式。MySQL 不支持这种方式。
- TABLE 表示通过表产生主键，即框架借助表模拟序列（Sequence）产生主键。使用该策略可以使应用更易于数据库的移植。

4. 映射 Blob 和 Clob 类型（@Lob）

在通常情况下，可以在数据库中保存图片、长文本等类型的数据。这些类型的数据一般会被保存成 Blob 和 Clob 类型。这两种类型的数据可以使用@Lob 来标注。例如，一个 RcJob 实体的字段：
```
@Lob
@Column(name = "job_comment")
private String jobComment; //工作说明
```
因为@Lob 标注的数据一般占用的内存空间较大，所以通常使用延迟加载的方式，一般与@Basic 标注同时使用，设置加载方式为 FetchType.LAZY，例如：
```
@Lob
@Column(name = "job_comment")
@Basic ( fetch = FetchType.LAZY)
private String jobComment;
```

5. 映射时间类型

在进行实体映射时，有关时间和日期的类型可以是 java.sql 包下的 java.sql.Date、java.sql.Time 或 java.sql.Timestamp，也可以是 java.util 包下的 java.util.Date 或 java.util.Calendar。在默认情况下，实体中使用的数据类型是 java.sql 包下的类，如果想使用 java.util 包下的时间和日期类型，则可以使用@Temporal 标注来说明。

在使用@Temporal 标注时，因为数据表对时间类型有更严格的要求，所以必须指定具体的时间类型。在 javax.persistence.TemporalType 枚举中定义了 3 种时间类型。

- DATE 对应 java.sql.Date。
- TIME 对应 java.sql.Time。
- TIMESTAMP 对应 java.sql.Timestamp。

例如：

```
@Column(name = "job_posttime")
@Temporal(TemporalType.DATE)
private Timestamp jobPosttime;
```

6. 瞬时字段（非持久化类型）

使用@Transient 标注不需要与数据库映射的字段（不需要保存到数据库中），例如：

```
@Transient
private int tempValue;  //临时用的属性，不需要保存到数据库中
```

7. 映射枚举类型

使用@Enumerated 标注枚举类型，例如：

```
@Column(name = "student_pltype")
@Enumerated (EnumType.STRING)
private PLType studentPltype;  //政治面貌
public enum PLType{
    MEMBER,PARTYMEMBER
}
```

3.3.2 实体关系映射

实体关系是指实体与实体之间的关系，需要从方向性和数量性两个方面来考虑。

两个实体间的关系从方向上分，可分为单向关联和双向关联。单向关联是在一个实体中引用了另一个实体。双向关联是两个实体之间可以相互获得对方对象的引用。

两个实体间的关系从引用的数量上分，可分为一对一（One to One）、一对多（One to Many）、多对一（Many to One）和多对多（Many to Many）关联。比较常用的就是一对多和多对一关联。

1. 一对一关联

在网上人才中心系统中，个人实体和用户实体存在一对一关联的关系。

在 rc_person 表中定义一个外键 user_id 用于与表 rc_user 关联，并给该外键添加唯一性约束，如图 3-6 所示。

第 3 章 基于 Spring Data JPA 实现数据访问层

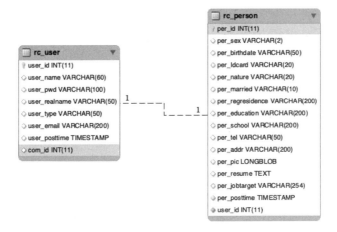

图 3-6 一对一关联的表

实体的定义如下：
```
@Entity
@Table(name = "rc_user")
public class RcUser{
    ...
}
@Entity
@Table(name = "rc_person")
public class RcPerson{
    ...
    @OneToOne（targetEntity = RcUser.class, optional = false）
    @JoinColumn(name = "user_id", referencedColumnName = "user_id", unique = true,nullable = false, updatable = false)
    private RcUser user;
    ...
}
```

@OneToOne 标注用于建立实体 Bean 之间的一对一关联。该标注具有以下属性。

- targetEntity 表示关联的实体类型，通常不需要设置，使用默认设置即可，但如果使用接口定义变量时，则需要使用该属性明确关联的类。
- cascade 表示与此实体一对一关联的实体的级联样式类型。级联样式是在对实体进行操作时，对关联表的操作策略，包括 CascadeType.ALL（所有操作）、CascadeType.MERGE（修改）、CascadeType.PERSIST（持久化）、CascadeType.REFRESH（刷新）和 CascadeType.REMOVE（删除）。在默认情况下，不关联任何操作。
- fetch 表示实体的加载方式，默认为即时加载 FetchType.EAGER，也可以使用延迟加载 FetchType.LAZY。
- optional 表示关联的实体对象的值是否允许为 null。默认为 true，表示允许为 null；如果将其设置为 false，则表示不允许为 null。
- mappedBy 用于双向关联实体时，标注在不保存关系的实体中，指明对方的相关属性。在建立单向关联时，不需要使用该属性。

@JoinColumn 标注用于表示关联的列，它的属性与@Column 标注类似。

上面只是建立了外键到主键的单向关联，如果要建立双向关联，则双方都需要添加映射。而且，主键需要使用属性 mappedBy。例如：

```
@Entity
@Table(name = "rc_user")
public class RcUser{
    ...
    @OneToOne(optional = false , cascade = CascadeType.ALL, mappedBy = "user", fetch=FetchType.LAZY)
    private RcPerson person;
    ...
}
```

2. 一对多关联

在网上人才中心系统中，如果我们希望在查询职位时，查看对该职位的申请，就可定义一对多关联。

如图 3-7 所示，表 rs_application 和表 rc_job 通过外键 job_id 实现一对多关联。

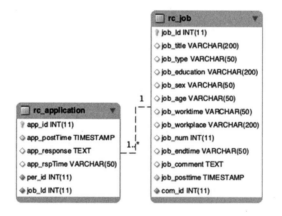

图 3-7　通过外键实现一对多关联的表

实体类的映射关系可以定义如下：

```
@Entity
@Table(name = "rc_job ")
public class RcJob{
    ...
    @OneToMany(targetEntity = BsDetails.class, fetch=FetchType.LAZY, cascade = {CascadeType.ALL})
    private List<RcApplication> applications;
}
```

@OneToMany 标注用于定义一对多关联，其属性包括：

- **targetEntity** 表示默认关联的实体类型。如果指定了集合泛型的具体类型，则可以省略该属性。

- cascade 表示与此实体一对多关联的实体的级联样式类型。级联样式是在对实体进行操作时，对关联表的操作策略，包括 CascadeType.ALL（所有操作）、CascadeType.MERGE（修改）、CascadeType.PERSIST（持久化）、CascadeType.REFRESH（刷新）、CascadeType.REMOVE（删除）。默认不级联。
- fetch 表示该实体的加载方式，默认为延迟加载 LAZY。
- mappedBy 用于双向关联实体时，标注在不保存关系的实体中（对应主表的实体）。如果类之间是单向关联，则不需要提供定义，如果类和类之间形成双向关联，我们就需要使用这个属性进行定义,否则可能引起数据一致性的问题。该属性的值是"多"方 class 中的"一"方的变量名。

3．多对一关联

例如，申请与职位之间可以定义多对一关联：

```
@Entity
@Table(name = "rc_application")
@Data
public class RcApplication{
    ...
    @ManyToOne(targetEntity = RcJob.class, cascade=CascadeType.REFRESH,optional=false)
    @JoinColoumn(name="job_id", , referencedColumnName = "job_id")
    private RcJob job;
    ...
}
```

@ManyToOne 标注用于定义多对一关联，其属性包括：
- targetEntity 表示关联的实体类型，通常不需要设置，使用默认设置即可，但如果使用接口定义变量时，则需要使用该属性明确关联的类。
- cascade 表示与此实体多对一关联的实体的级联样式类型。
- fetch 表示实体的加载方式，默认为即时加载 FetchType.EAGER，也可以使用延迟加载 FetchType.LAZY。
- optional 表示关联的实体对象的值是否允许为 null。默认为 true，表示允许为 null；如果将其设置为 false，则表示不允许为 null。

@JoinColumn 标注用于指定与所操作实体或实体集合相关的数据库表中的列字段。其属性包括：
- name 表示对应的字段。
- referencedColumnName 表示关联的字段。

也可以建立双向关联。在一端使用属性 mappedBy 表示一端是关系被维护端，多端是关系维护端（inverse side）。在建表时，应在关系维护端建立外键列指向关系被维护端的主键。例如：

```
public class RcJob {
    ...
    @OneToMany(targetEntity = RcApplication.class, cascade = CascadeType.ALL, mappedBy = "job")
    private List<RcApplication> applications;
    ...
}
```

4. 多对多关联

多对多关联一般是通过中间表实现的。如图 3-8 所示，用户表 rc_user 和角色表 rc_role 之间是多对多关联的关系，中间表是 rc_user_role。

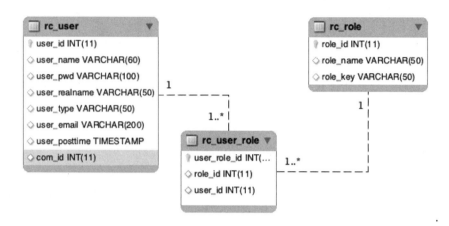

图 3-8　通过中间表实现多对多关联的表

管理员实体和角色实体之间可以建立单向多对多关联。例如：

```
@Entity
@Table(name="rc_user")
public class RcUser{
    ...
    @ManyToMany(cascade = CascadeType.All, fetch = FetchType.LAZY)
    @JoinTable(name = "rc_user_role",joinColumns = @JoinColumn(name="user_id",referencedColumnName = "user_id"),inverseJoinColumns = @JoinColumn(name = " role_id",referencedColumnName = "role_id"))
    private List<RcRole> roles;
    ...
}
@Entity
@Table(name="rc_role")
public class RcRole{
    ...
}
```

角色和权限之间也是多对多关系，可以建立双向多对多关联。例如：

```
@Entity
@Table(name="rc_role")
public class RcRole{
    ...
    @ManyToMany(cascade = CascadeType.All, fetch = FetchType.LAZY)
    @JoinTable(name = "rc_role_person",joinColumns = @JoinColumn(name="role_id",referencedColumnName = "role_id"),inverseJoinColumns = @JoinColumn(name = " perm_id",referencedColumnName = "perm_id"))
    private List<RcPermission> permissions;
    ...
}
@Entity
```

```
@Table(name="rc_permission")
public class RcPermission{
    ...
    //多对多关联
    @ManyToMany(mappedBy="categories")   //放弃维护关系，由 role 维护
    private List<RcRole> roles;
    ...
}
```

3.3.3　使用逆向工程生成实体类

如果用户习惯面向对象的思维方式，则可以先设计实体类并配置好映射。同时，在配置文件中设置 spring.jpa.hibernate.ddl-auto=update。在程序启动后，就可以自动创建数据表。

但是如果用户习惯先设计数据库，也可以采用逆向工程的方式生成实体类。具体方法如下所述。

（1）选择【View】→【Tool Windows】→【Persistence】命令，如图 3-9 所示，打开持久化工具窗口，如图 3-10 所示。

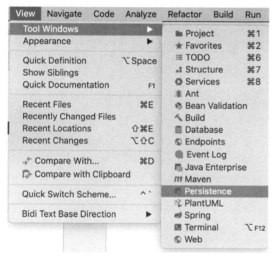

图 3-9　选择【Persistence】命令　　　　图 3-10　持久化工具窗口

（2）在持久化工具窗口的项目名称上右击，并在弹出的快捷菜单中选择【Generate Persistence Mapping】→【By Database Schema】命令，如图 3-11 所示。

（3）在如图 3-12 所示的界面中，选择数据源和实体类所存放的包，选择除中间表以外的所有表，其他选项按图 3-12 进行设置，单击【OK】按钮，即可生成实体类。这里取消了生成关联关系，可以在后面根据需要自行配置。此外，自动增值的属性需要添加如下内容：

```
@GeneratedValue(strategy=GenerationType.IDENTITY)
```

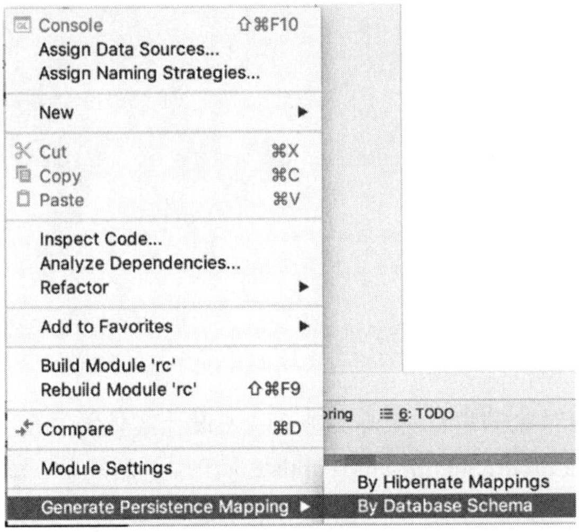

图 3-11 选择【By Database Schema】命令

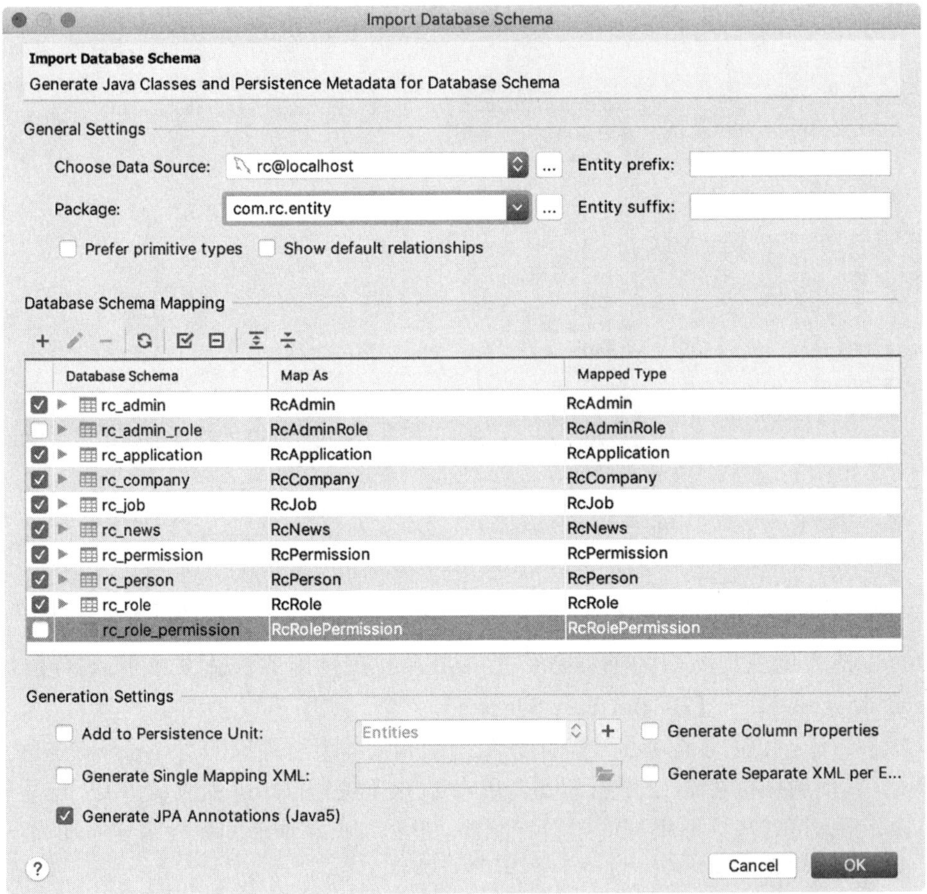

图 3-12 生成实体类

3.3.4 网上人才中心系统实体类定义

下面的实体类定义使用了 Lombok 的@Data 标注。对于双向关联而言,为了避免 Lombok 自动生成 equals()和 hashCode,引用了关联实体,产生循环关联,并使用了@EqualsAndHashCode 标注,通过 exclude 排除了关联属性。

1. 用户实体类

```
@Entity
@Table(name = "rc_user", schema = "rc")
@Data
public class RcUser{
    @Id
    @Column(name = "user_id", nullable = false)
    @GeneratedValue(strategy=GenerationType.IDENTITY)
    private Integer userId;
    @Column(name = "user_name", nullable = true, length = 50)
    private String userName;
    @Column(name = "user_pwd", nullable = true, length = 50)
    private String userPwd;
    @Column(name = "user_type", nullable = true, length = 45)
    private String userType;
    @Column(name = "user_email", nullable = true, length = 150)
    private String userEmail;
    @Column(name = "user_posttime")
    @Temporal(TemporalType.DATE)
    private Date userPosttime;
    @ManyToMany(fetch = FetchType.LAZY)
    @JoinTable(name = "rc_admin_role",joinColumns = @JoinColumn(name="user_id",
    referencedColumnName = "user_id"),inverseJoinColumns = @JoinColumn(name = "role_id",
    referencedColumnName = "role_id"))
    private List<RcRole> roles;
}
```

2. 企业实体类

```
@Entity
@Table(name = "rc_company", schema = "rc")
@Data
public class RcCompany {
    @Id
    @Column(name = "com_id", nullable = false)
    @GeneratedValue(strategy=GenerationType.IDENTITY)
    private Integer comId;
    @Column(name = "com_name", nullable = true, length = 200)
    private String comName;
    @Column(name = "com_type", nullable = true, length = 50)
    private String comType;
    @Column(name = "com_createtime", nullable = true, length = 50)
    private String comCreatetime;
```

```
    @Column(name = "com_contactor", nullable = true, length = 50)
    private String comContactor;
    @Column(name = "com_empnum", nullable = true)
    private Integer comEmpnum;
    @Column(name = "com_tel", nullable = true, length = 50)
    private String comTel;
    @Column(name = "com_fax", nullable = true, length = 50)
    private String comFax;
    @Column(name = "com_addr", nullable = true, length = 200)
    private String comAddr;
    @Column(name = "com_zip", nullable = true, length = 10)
    private String comZip;
    @Column(name = "com_homepage", nullable = true, length = 200)
    private String comHomepage;
    @Column(name = "com_email", nullable = true, length = 200)
    private String comEmail;
    @Column(name = "com_desc", nullable = true, length = -1)
    private String comDesc;
    @Column(name = "com_posttime", nullable = false)
    private Timestamp comPosttime;
}
```

3. 个人实体类

```
@Entity
@Table(name = "rc_person", schema = "rc")
@Data
public class RcPerson {
    @Id
    @Column(name = "per_id", nullable = false)
    @GeneratedValue(strategy=GenerationType.IDENTITY)
    private Integer perId;
    @OneToOne（targetEntity = RcUser.class, optional = false）
    @JoinColumn(name = "user_id", referencedColumnName = "user_id", unique = true,nullable = false, updatable = false)
    private RcUser user;

    @Column(name = "per_sex", nullable = true, length = 2)
    private String perSex;
    @Column(name = "per_birthdate", nullable = true, length = 50)
    private String perBirthdate;
    @Column(name = "per_Idcard", nullable = true, length = 20)
    private String perIdcard;
    @Column(name = "per_nature", nullable = true, length = 20)
    private String perNature;
    @Column(name = "per_married", nullable = true, length = 10)
    private String perMarried;
    @Column(name = "per_regresidence", nullable = true, length = 200)
    private String perRegresidence;
    @Column(name = "per_education", nullable = true, length = 200)
```

```java
    private String perEducation;
    @Column(name = "per_school", nullable = true, length = 200)
    private String perSchool;
    @Column(name = "per_tel", nullable = true, length = 50)
    private String perTel;
    @Column(name = "per_email", nullable = true, length = 200)
    private String perEmail;
    @Column(name = "per_addr", nullable = true, length = 200)
    private String perAddr;
    @Column(name = "per_pic", nullable = true, length = 200)
    private String perPic;
    @Column(name = "per_resume", nullable = true, length = -1)
    private String perResume;
    @Column(name = "per_jobtarget", nullable = true, length = 254)
    private String perJobtarget;
    @Column(name = "per_posttime", nullable = false)
    private Timestamp perPosttime;
}
```

4．职位实体类

```java
@Entity
@Table(name = "rc_job", schema = "rc")
@Data
public class RcJob {
    @Id
    @Column(name = "job_Id", nullable = false)
    @GeneratedValue(strategy=GenerationType.IDENTITY)
    private Integer jobId;
    @Column(name = "job_title", nullable = true, length = 200)
    private String jobTitle;
    @Column(name = "job_type", nullable = true, length = 50)
    private String jobType;
    @Column(name = "job_education", nullable = true, length = 200)
    private String jobEducation;
    @Column(name = "job_sex", nullable = true, length = 50)
    private String jobSex;
    @Column(name = "job_age", nullable = true, length = 50)
    private String jobAge;
    @Column(name = "job_worktime", nullable = true, length = 50)
    private String jobWorktime;
    @Column(name = "job_workplace", nullable = true, length = 200)
    private String jobWorkplace;
    @Column(name = "job_num", nullable = true)
    private Integer jobNum;
    @Column(name = "job_endtime", nullable = true, length = 50)
    private String jobEndtime;
    @Column(name = "job_comment", nullable = true, length = -1)
    private String jobComment;
    @Column(name = "job_posttime", nullable = false)
```

```java
    private Timestamp jobPosttime;
    @ManyToOne(optional=false)
    @JoinColumn(name="com_id", referencedColumnName = "user_id")
    private RcCompany company;
}
```

5. 申请实体类

```java
@Entity
@Table(name = "rc_application", schema = "rc")
@Data
public class RcApplication {
    @Id
    @Column(name = "app_id", nullable = false)
    @GeneratedValue(strategy=GenerationType.IDENTITY)
    private Integer appId;
    @Column(name = "app_postTime", nullable = false)
    private Timestamp appPostTime;
    @Column(name = "app_response", nullable = true, length = -1)
    private String appResponse;
    @Column(name = "app_rspTime", nullable = true, length = 50)
    private String appRspTime;
    @ManyToOne(optional=false)
    @JoinColumn(name="per_id", referencedColumnName = "user_id")
    private RcPerson person;
    @ManyToOne(optional=false)
    @JoinColumn(name="job_id", referencedColumnName = "job_id")
    private RcJob job;
}
```

6. 角色实体类

```java
@Entity
@Table(name = "rc_role", schema = "rc")
@Data
@EqualsAndHashCode(exclude = {"permissions"})
public class RcRole {
    @Id
    @Column(name = "role_id", nullable = false)
    @GeneratedValue(strategy = GenerationType.IDENTITY)
    private int roleId;
    @Column(name = "role_name", nullable = true, length = 50)
    private String roleName;
    @Column(name = "role_key", nullable = true, length = 50)
    private String roleKey;
    @ManyToMany(fetch = FetchType.LAZY)
    @JoinTable(name = "rc_role_permission", joinColumns = @JoinColumn(name = "role_id", referencedColumnName = "role_id"), inverseJoinColumns = @JoinColumn(name = "perm_id", referencedColumnName = "perm_id"))
    private List<RcPermission> permissions;
}
```

6．权限实体类

```
@Entity
@Table(name = "rc_permission", schema = "rc")
@Data
@EqualsAndHashCode(exclude = {"roles"})
public class RcPermission {
    @Id
    @Column(name = "perm_id", nullable = false)
    @GeneratedValue(strategy=GenerationType.IDENTITY)
    private Integer permId;
    @Column(name = "perm_name", nullable = true, length = 50)
    private String permName;
    @Column(name = "perm_name", nullable = true, length = 50)
    private String permUrl;
    @Column(name = "perm_name", nullable = true, length = 50)
    private List<RcRole> roles;
}
```

7．新闻实体类

```
@Entity
@Table(name = "rc_news", schema = "rc")
@Data
public class RcNews {
    @Id
    @Column(name = "news_id", nullable = false)
    @GeneratedValue(strategy=GenerationType.IDENTITY)
    private Integer newsId;
    @Column(name = "news_title", nullable = false, length = 200)
    private String newsTitle;
    @Column(name = "news_content", nullable = false, length = -1)
    private String newsContent;
    @Column(name = "news_postTime", nullable = false)
    private Timestamp newsPostTime;
}
```

3.4 JPA 数据操作方法

3.4.1 使用预定义的方法查询

我们在定义自己的 Dao 接口时，使其继承 JpaRepository 接口，就可以获得 Spring 预先定义的多种数据操作方法。

1．基本查询方法

基本查询方法可以完成基本的增、删、改、查操作。常用的方法包括：

- <S extends T> S save(S entity);//保存
- List<T> findAll();//查询所有对象

- List<T> findAll(Sort var1);//查询所有对象并排序
- void deleteById(ID var1);//按 ID 删除
- Page<T> findAll(Pageable var1);//查询所有对象,分页并排序
- Optional<T> findById(ID var1);//按 ID 查询

例如：

```
userDao.save(user);                                  //添加或保存用户
userDao.deleteById(userId);                          //按 ID 删除用户
RcUser user=userDao.findById(userId).get();          //按 ID 查找管理员
Sort sort = Sort.by(Sort.Direction.DESC, "appPosttime");
Page<RcApplication> page=appDao.findAll(sort); //查询申请并排序
Sort sort1 = Sort.by(Sort.Direction.DESC, "jobPosttime");
Pageable pageable = PageRequest.of(pageNo-1, pageSize,sort1);
Page<RcJob> page=jobDao.findAll(pageable);    //查询职位,分页并排序
```

2. 根据实例查询

根据实例查询就是通过一个例子来查询。即查询条件也是一个实体对象，以一个现有的客户对象为例，查询和这个客户对象相匹配的对象。常用的方法包括：

- <S extends T> List<S> findAll(Example<S> var1);//根据实例查询对象
- <S extends T> List<S> findAll(Example<S> var1, Sort var2);//根据实例查询对象并排序
- <S extends T> Page<S> findAll(Example<S> var1, Pageable var2);//根据实例查询对象

例如，现在要查询地址是河南省郑州市且姓刘的客户，代码如下：

```
RcPerson person = new RcPerson();                    //创建查询条件,即数据对象
person.setPerRealname("刘");
customer.setPerAddr ("河南省郑州市");
ExampleMatcher matcher = ExampleMatcher.matching()   //构建匹配器,决定如何使用条件
    .withMatcher("perRealname", GenericPropertyMatchers.startsWith());//姓名采用开始匹配的方式查询
Example<RcPerson> ep = Example.of(customer, matcher); //创建实例
List<RcPerson> ls = personDao.findAll(ep);           //查询
```

在上面的代码中，创建了一个 RcPerson 对象，即一个实例，代表查询条件。ExampleMatcher 对象是匹配器，表示如何使用实体对象中的值进行查询，代表查询方式，用于解释如何查询。Example 对象将 RcPerson 对象和 ExampleMatcher 对象包装在一起，可以根据它进行查询。

对于简单的查询，也可以不使用匹配器。例如，查询地址是北京市且性别为男的人才，代码如下：

```
RcPerson person= new RcPerson();
person.setPerAddr("北京市");
person.setPerSex("男");
Example<RcPerson> ex = Example.of(person);    //创建实例
List<Customer> ls = dao.findAll(ex);          //查询
```

根据实例，也可以进行模糊查询，例如，根据地址、简介进行模糊查询，可忽略英文字母的大小写，但要求地址采用开始匹配的方式查询，代码如下：

```
RcPerson person= new RcPerson();
```

```
person.setPerAddr("河北省石家庄");
customer.setResume("BB");
//创建匹配器，即如何使用查询条件
ExampleMatcher matcher = ExampleMatcher.matching()         //构建对象
    .withStringMatcher(StringMatcher.CONTAINING)   //改变默认字符串匹配方式为模糊查询
    .withIgnoreCase(true)                          //改变默认大小写忽略方式：忽略大小写
    .withMatcher("perAddr", GenericPropertyMatchers.startsWith())//地址采用开始匹配的方式查询
Example<RcPerson> ex = Example.of(person);         //创建实例
List<RcPerson> ls = personDao.findAll(ex);         //查询
```

需要注意的是，使用这种查询方法不支持过滤条件分组，即不支持过滤条件使用 or（或）来连接，所有的过滤条件都简单地使用 and（并且）来连接。在进行实例查询时，若实例对象中的某属性没有输入值，就忽略这个过滤条件。

3．复杂查询

在某些情况下，查询条件会非常多，需要不断地拼接属性，这时就需要使用 JpaSpecificationExecutor 接口了。

- Root root 代表可以查询和操作的实体对象的根，可以通过 get("属性名")方法来获取对应的值。
- CriteriaQuery query 代表一个 specific 的顶层查询对象，它包含查询的各个部分，如 select、from、where、group by、order by 等。
- CriteriaBuilder cb 代表构建 CriteriaQuery 的构建器对象，相当于条件或条件组合，并以 Predicate 的形式返回。

例如，单条件查询代码如下：

```
@Test
public void test1(){
    Specification<RcUser> spec = new Specification<RcUser>() {
        @Override
        public Predicate toPredicate(Root<RcUser> root,
                                    CriteriaQuery<?> query, CriteriaBuilder cb) {
            Predicate pre = cb.equal(root.get("userRealname"), "王五");
            return pre;
        }
    };
    List<RcUser> list = userDao.findAll(spec);
    for (RcUser user : list) {
        System.out.println(user.getUserRealname());
    }
}
```

多条件查询代码如下：

```
@Test
public void test2() {
    Specification<RcUser> spec = new Specification<RcUser>() {
        @Override
        public Predicate toPredicate(Root<RcUser> root,
                                    CriteriaQuery<?> query, CriteriaBuilder cb) {
```

```
            List<Predicate> list = new ArrayList<>();
            list.add(cb.equal(root.get("userRealname"), "王五"));
            list.add(cb.equal(root.get("userType"),"企业用户"));
            //此时条件之间是没有任何关系的
            Predicate[] arr = new Predicate[list.size()];
            return cb.and(list.toArray(arr));
        }
    };
    List<RcUser> list = usersDao.findAll(spec);
    for (RcUser user: list) {
        System.out.println(user.getUserRealname());
    }
}
```

3.4.2 使用自定义方法查询

1. 通过自动解析查询

按照 Spring Data JPA 定义的规则，查询方法以 find、read、get 开头（如 find、findBy、read、readBy、get、getBy）。在涉及条件查询时，条件的各属性使用条件关键字连接。需要注意的是，条件属性的首字母需大写。框架在进行方法名解析时，会先把方法名中多余的前缀删除，再对剩余的部分进行解析。如果方法的最后一个参数是 Sort 或 Pageable 类型，则框架会提取相关的信息，以便按照规则进行排序或分页查询。方法解析的语法样例如表 3-1 所示。

表 3-1 方法解析的语法样例

Keyword	Sample	JPQL snippet
And	findByLastnameAndFirstname	… where x.lastname = ?1 and x.firstname = ?2
Or	findByLastnameOrFirstname	… where x.lastname = ?1 or x.firstname = ?2
Is, Equals	findByFirstname, findByFirstnameIs, findByFirstnameEquals	… where x.firstname = ?1
Between	findByStartDateBetween	… where x.startDate between ?1 and ?2
LessThan	findByAgeLessThan	… where x.age < ?1
LessThanEqual	findByAgeLessThanEqual	… where x.age <= ?1
GreaterThan	findByAgeGreaterThan	… where x.age > ?1
GreaterThanEqual	findByAgeGreaterThanEqual	… where x.age >= ?1
After	findByStartDateAfter	… where x.startDate > ?1
Before	findByStartDateBefore	… where x.startDate < ?1
IsNull	findByAgeIsNull	… where x.age is null
IsNotNull, NotNull	findByAge(Is)NotNull	… where x.age not null
Like	findByFirstnameLike	… where x.firstname like ?1

续表

Keyword	Sample	JPQL snippet
NotLike	findByFirstnameNotLike	… where x.firstname not like ?1
StartingWith	findByFirstnameStartingWith	… where x.firstname like ?1 (parameter bound with appended %)
EndingWith	findByFirstnameEndingWith	… where x.firstname like ?1 (parameter bound with prepended %)
Containing	findByFirstnameContaining	… where x.firstname like ?1 (parameter bound wrapped in %)
OrderBy	findByAgeOrderByLastnameDesc	… where x.age = ?1 order by x.lastname desc
Not	findByLastnameNot	… where x.lastname <> ?1
In	findByAgeIn(Collection<Age> ages)	… where x.age in ?1
NotIn	findByAgeNotIn(Collection<Age> ages)	… where x.age not in ?1
True	findByActiveTrue()	… where x.active = true
False	findByActiveFalse()	… where x.active = false
IgnoreCase	findByFirstnameIgnoreCase	… where UPPER(x.firstame) = UPPER(?1)

比如，我们对申请（RcApplication）进行查询。方法名为 findByJobJobTitle。解析的过程如下所述。

（1）首先删除 findBy，然后对剩余的属性进行解析。

（2）判断 jobJobTitle（首字母变为小写后）是否为 RcApplication 的一个属性，如果是，则表示根据该属性进行查询；否则，继续进行步骤 3。

（3）从右向左截取第一个大写字母开头的字符串（此处为 Title），然后检查剩余的字符串 jobJob（首字母变小写后）是否为 RcApplication 的一个属性，如果是，则表示根据该属性进行查询；否则，重复步骤 3，继续从右向左截取；最后假设 job 为 RcApplication 的一个属性。

（4）继续处理剩余的部分（JobTitle），先判断 job 所对应的类型是否有 jobTitle 属性，如果有，则表示该方法最终是根据"job.jobTitle"的取值进行查询的；否则继续按照步骤 3 的规则从右向左截取。

2．自定义查询

在 JPA 中，可以执行两种方式的查询：一种是使用 JPQL；另一种是使用 Native SQL（本地 SQL）。JPQL 是一种与数据库无关的，基于实体（entity-based）的查询语言。使用 JPQL 查询可以消除不同数据库的差异，便于在各种数据库之间进行切换。JPQL 和普通的 SQL 很相似，但它是基于实体的查询。例如：

select u from RcUser u where u.userName=? and u.userPwd=?

这里使用的 RcPerson 是类，因此需要区分大小写。p.userName 和 p.userPwd 使用的是对象属性，而不是表中的字段，因此也需要区分大小写。

在定义查询时，使用@Query 标注。@Query 标注的使用非常简单，只需在声明的方法上加上该标注，同时提供一个 JPQL 查询语句即可。例如：

```
@Query("select u from RcUser u where u.userEmail = ?1")
RcUser selectByEmail(String email);
@Query("select u from RcUser p where u.userName = ?1 and u.userPwd= ?2")
RcUser selectByNameAndPwd(String userName, String userPwd);
@Query("select u from RcUser u where u.userName like %?1%")
List<RcUser> selectByUserNameLike(String userName);
```

在默认情况下,Spring Data JPA 使用基于位置的参数绑定,如前文所有示例所述。这使得查询方法在重构参数位置时容易出错。若要解决此问题,可以使用@Param 标注为方法参数指定具体名称并在查询中绑定名称。例如:

```
@Query("select u from RcUser u where u.userName = :userName")
RcUser getByUserName(@Param("userName") String userName)
```

在使用@Query 标注涉及修改、删除操作时,需要加上@Modifying 标注。例如:

```
@Transactional()
@Modifying
@Query("delete RcUser p where p.userId = ?1")
int deleteByUserId(Integer userId);
```

@Query 标注支持将 nativeQuery 设置为 true 来执行原生查询,还支持基于位置的参数绑定及参数命名。例如:

```
@Query(value = "select * from rc_user p where p.user_name = ?1", nativeQuery = true)
RcUser selectByUserName(String userName);
```

3.4.3 查询结果格式

除了前文提到的查询一个实体或多个实体列表,还有如下所述的一些情况。

1. Map-List

列表的元素为 Map 类型。例如:

```
@Query("select new map(t.name as name,t.value as value) from Test t where t.name = :keyWord")
List<Map<String,Object>> findTest(@Param("keyWord") String keyWord);
```

注意:其中的 as name 必须写,若不写,则 Map 中的 key 为下标。

2. Vo-List

查询部分字段,并通过构造函数放到对象里。例如:

```
@Query("select new TestVo(t.name as name,t.value as value) from Test t where t.name = :keyWord")
List<TestVo>findTest(@Param("keyWord") String keyWord);
```

TestVo 的定义如下:

```
public class TestVo{
    private String name;
    private String value;
    public TestVo(String name,String value){
        this.name = name;
        this.value = value;
    }
    // get..set..
}
```

注意：在使用构造方法的形式进行初始化时，传入顺序必须与构造函数顺序一致。

3．Projection-List

查询结果被放到接口列表里。例如：
@Query("select t.name as name,t.value as value from Test t where t.name = :keyWord")
List<TestProjection>findTest(@Param("keyWord") String keyWord);
接口定义如下：
public interface TestProjection{
　　String getName();
　　String getValue();
}

注意：TestProjection 为接口，有对应的 get()方法，这种模式不需要使用 new 的方式传值。

3.4.4　网上人才中心系统数据访问层设计

在数据访问层中，我们定义了一个基接口 RcBaseDao，该接口继承于 JpaRepositoryImplementation 接口，并新增了一个方法，用于查询最新添加的前 num 个记录，这个接口的定义已经在 3.2.4 节中给出。RcBaseDao 接口有一个实现类 RcBaseDaoImpl，该类继承于 SimpleJpaRepository 类。数据访问层的其他接口继承于 RcBaseDao 接口。数据访问层相关接口和类如图 3-13 所示，反映了其与原有 Spring Data JPA 接口和类的关系。

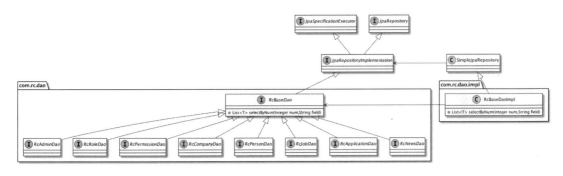

图 3-13　数据访问层相关接口和类

3.4.5　网上人才中心系统业务逻辑层设计

由于数据访问层的实现方式有所变化，因此网上人才中心系统的业务逻辑层也需要进行相应调整。这里以企业业务逻辑层为例，代码如下：
@Service
public class RcCompanyServiceImpl implements RcCompanyService {
　　@Resource
　　private RcCompanyDao companyDao;

　　@Override

```java
public void addCompany(RcCompany com) {
    companyDao.save(com);
}

@Override
public void editCompany(RcCompany com) {
    companyDao.save(com);
}

@Override
public void deleteCompany(Integer userId) {
    companyDao.deleteById(userId);
}

@Override
public RcCompany findCompanyById(Integer userId) {
    return companyDao.findById(userId).get();
}

@Override
public RcCompany findCompany(String userName, String userPwd) {
    return companyDao.findByUserNameAndUserPwd(userName,userPwd);
}

@Override
public List<RcCompany> findCompanies(Integer num) {
    Sort sort = Sort.by(Sort.Direction.DESC, "comPosttime");
    Pageable pageable = PageRequest.of(0, num, sort);
    return companyDao.findAll(pageable).toList();
}

@Override
public Page<RcCompany> findCompanies(String comType, String comName, Integer pageNo, Integer pageSize) {
    Specification<RcCompany> spec = new Specification<RcCompany>() {
        @Override
        public Predicate toPredicate(Root<RcCompany> root, CriteriaQuery<?> criteriaQuery, CriteriaBuilder criteriaBuilder) {
            List<Predicate> list = new ArrayList<>();
            if (comType != null) list.add(criteriaBuilder.like(root.get("comType"), "%"+comType+"%"));
            if (comName != null) list.add(criteriaBuilder.like(root.get("comName"), "%"+comName+"%"));
            Predicate[] arr = new Predicate[list.size()];
            return criteriaBuilder.and(list.toArray(arr));
        }
    };
    if(pageNo==null)pageNo=1;
    if(pageSize==null)pageSize=10;
```

```java
        Sort sort = Sort.by(Sort.Direction.DESC, "comPosttime");
        Pageable pageable = PageRequest.of(pageNo - 1, pageSize, sort);
        return companyDao.findAll(spec, pageable);
    }
}
```

第 4 章　基于 MyBatis 实现数据访问层

本章主要介绍 MyBatis 技术，Spring Boot 与 MyBatis 整合的方法，以及 MyBatis Generator 和 MyBatis-Plus 两种逆向工程技术。

4.1　MyBatis 技术概述

4.1.1　MyBatis 简介

MyBatis 原本是 Apache 的一个开源项目 iBatis。2010 年，这个项目由 Apache Software Foundation 迁移到了 Google Code，并且改名为 MyBatis。2013 年 11 月，这个项目又迁移到了 GitHub。

MyBatis 是一款优秀的持久层框架，它支持定制化 SQL、存储过程及高级映射。MyBatis 避免了几乎所有的 JDBC 代码编写、手动设置参数及结果集的检索。MyBatis 可以使用简单的 XML 或标注来配置和映射原始信息，将接口和 Java 的 POJOs（Plain Ordinary Java Objects，普通的 Java 对象）映射成数据库中的记录。

MyBatis 具有以下特点。

- 简单易学：本身体积就很小且简单；没有任何第三方依赖，最简单的安装只需要两个 jar 文件，并配置几个 SQL 映射文件；易于学习，易于使用，通过文档和源代码可以比较完全地掌握它的设计思路和实现方法。
- 灵活：MyBatis 不会对应用程序或数据库的现有设计产生任何影响。SQL 语句被编写在 XML 文件里，便于统一管理和优化。通过 SQL 语句可以满足操作数据库的所有需求。
- 解除 SQL 语句与程序代码的耦合：通过提供 Dao 层将业务逻辑和数据访问逻辑分离，使系统的设计更清晰、更容易维护、更容易进行单元测试。SQL 语句和代码的分离，提高了可维护性。
- 提供映射标注，支持对象与数据库的 ORM 字段的关系映射。
- 提供对象关系映射标签，支持对象关系组件维护。
- 提供 XML 标签，支持编写动态 SQL 语句。

4.1.2　MyBatis 与 Spring Data JPA 比较

市场上主流的 ORM 技术主要涉及 Hibernate、JPA、Spring Data JPA 及 MyBatis。

Spring Data JPA 可以被理解为 JPA 规范的再次封装抽象，其底层仍然使用了 Hibernate 的 JPA 技术实现。所以，使用 Spring Data JPA 一般就会使用 Hibernate。这样看来，Spring Data JPA 与 MyBatis 对比，其实主要就是 Hibernate 与 MyBatis 对比。

Hibernate 是全自动的，而 MyBatis 是半自动的。Hibernate 完全可以通过对象关系模型实现对数据库的操作，拥有完整的 JavaBean 对象与数据库的映射结构，可以自动生成 SQL 语句。而 MyBatis 仅拥有基本的字段映射，对象数据及对象实际关系仍然需要通过手写 SQL 语句来实现和管理。

Hibernate 经过 JPA 规范的再次封装，功能强大，数据库无关性好，O/R 映射能力强，持久化层代码更加简单，开发速度很快。Hibernate 更适合使用面向对象的思维方式，使开发者无须关心 SQL 语句的生成与结果映射，可以更专注于业务流程。因此 Hibernate 有更好的缓存机制和较好的移植性，但其学习门槛高，而且如何设计 O/R 映射，如何在性能和对象模型之间取得平衡，以及如何用好 Hibernate 等要求用户的经验丰富且能力较强。

MyBatis 需要手动编写 SQL 语句，可以进行更为细致的 SQL 优化，可以减少查询字段。对于复杂的业务而言，MyBatis 的灵活性更大，持久化层的运行性能会更好。MyBatis 入门简单，即学即用，提供了数据库查询的自动对象绑定功能，而且延续了很好的 SQL 语句使用经验，对于对对象模型要求不高的项目来说，相当完美。

从深层次来看，对于数据的操作而言，Hibernate 是面向对象的，而 MyBatis 是面向关系的。面向对象考虑的是对象的整个生命周期，包括对象的创建、持久化、状态的改变和行为等，其中，对象的持久化只是对象的一种状态，而面向关系则更关注数据的高效存储和读取。

在实际应用中，需要根据不同的项目需求选择不同的框架。在框架的使用过程中，也要考虑框架的优势和劣势，扬长避短，发挥出框架的最大效用，才能真正地提高项目研发效率、完成项目的目标。

4.1.3 MyBatis 核心类及工作原理

1. MyBatis 核心类

MyBatis 核心类包括以下几种。

- SqlSessionFactoryBuilder（会话工厂创建类）：此类用于读取配置文件，创建 SqlSessionFactory 实例，其生命周期只涉及初始化阶段。XML 配置文件中包含了对 MyBatis 系统的核心设置，包括获取数据库连接实例的数据源（DataSource），以及决定事务作用域和控制方式的事务管理器（TransactionManager）。
- SqlSessionFactory（会话工厂类）：此类是 MyBatis 最基础的类，用来创建会话（即 SqlSession 实例），其生命周期与整个系统的生命周期相同。在系统运行的任何时候都可以使用它查询当前数据库的配置信息等。
- SqlSession（会话类）：此类包含了面向数据库执行 SQL 语句所需的所有方法，用于直接执行已映射的 SQL 语句。SqlSession 类对应一次数据库会话，所以在每次访问数据库时都需要在 SqlSessionFactory 实例的 openSession()方法中创建它。

- Executor（执行器类）：在 MyBatis 底层定义了一个 Executor 接口，用来操作数据库。它可以根据 SqlSession 实例传递的参数动态地生成需要执行的 SQL 语句，同时负责对查询缓存进行维护。
- MappedStatement（映射语句类）：在 Executor 接口操作数据库的方法中，包括一个 MappedStatement 类型的参数，该参数是对映射信息的封装，用于存储要映射的 SQL 语句的 ID、参数等信息。

2．MyBatis 工作过程

当框架启动时，MyBatis 工作过程如图 4-1 所示。

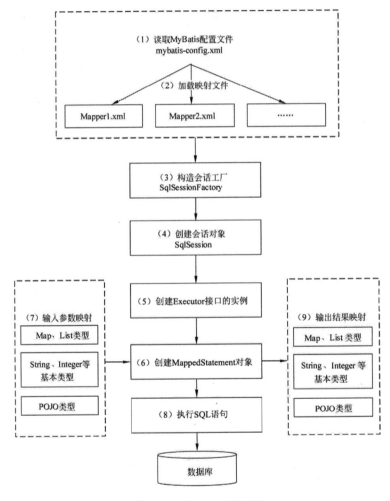

图 4-1　MyBatis 工作过程

（1）读取配置文件。SqlSessionFactoryBuilder 类利用 XMLConfigBuilder 解析配置文件，生成 Configuration 对象。

（2）加载映射文件。加载并解析映射文件，将其包装成 MappedStatement 对象，并将 MappedStatement 对象注册到全局的 Configuration 对象上。

（3）创建会话工厂。利用 Configuration 对象创建 SqlSessionFactory 对象。

（4）创建会话对象。利用 SqlSessionFactory 对象创建 SqlSession 对象。

（5）创建 Executor 接口的实例。在 MyBatis 底层定义一个 Executor 接口，用来操作数据库。它可以根据 SqlSession 传递的参数动态生成需要执行的 SQL 语句，同时负责对查询缓存进行维护。MyBatis 给 Executor 接口提供了多个实现类。

（6）创建 MappedStatement 对象。在 Executor 接口操作数据库的方法中，包括一个 MappedStatement 类型的参数，此参数封装了映射信息，一个 SQL 元素对应一个 MappedStatement 对象，SQL 元素的 ID 就是 MappedStatement 的 ID。

（7）输入参数映射。将输入的参数（基本类型、POJO 类型、List 类型、Map 类型）映射到 SQL 语句中。

（8）执行 SQL 语句。将最终得到的 SQL 语句和参数提交到数据库中执行，得到操作数据库的结果。

（9）输出结果映射。将操作数据库的结果按照映射的配置进行转换，可以转换为 HashMap、JavaBean 或者基本数据类型，并将最终结果返回。

4.1.4　映射器与 Mapper 实例

1．映射器

映射器是 MyBatis 中核心的组件之一，所有数据库的操作都会在映射器中被定义。也就是说，我们对数据库的 CRUD 操作，都是通过映射器实现的。映射器描述了要执行的每条 SQL 语句，每条语句的参数和返回值类型，结果集与实体属性的映射关系，以及缓存等信息。

有了映射器，就可以通过 SqlSession 发送 SQL 语句了。我们以 selectById 语句为例来了解如何发送 SQL 语句。比如，有一个用户表 test_user，其中包括两个字段 user_id、user_name，实体类是 User。一个简单的映射器的代码如下：

```xml
<?xml version="1.0" encoding="UTF-8"?>
<!DOCTYPE mapper PUBLIC "-//mybatis.org//DTD Mapper 3.0//EN"
        "http://mybatis.org/dtd/mybatis-3-mapper.dtd">
<mapper namespace="com.test.dao.UserDao">
    <resultMap type="com.test.entity.User" id="userResult" autoMapping="false">
        <id property="userId" column="user_id"/>
        <result property="userName" column="user_name"/>
    </resultMap>
    <select id="selectById" resultMap="userResult" parameterType="java.lang.Integer">
        select * from test_user where user_id = #{userId}
    </select>
</mapper>
```

上述映射器中包含了一个按 ID 查询的 SQL 配置 selectById 和结果集映射配置 userResult。

有了映射器，就可以使用 SqlSession 来执行查询操作。例如：

```
User user = (User)SqlSession.select("com.test.dao.UserDao.selectById ",3);
```

上述操作的第一个参数是由一个命名空间加上 SQL ID 组合而成的，可以完全定位一条 SQL 语句，这样 MyBatis 就会找到对应的 SQL 语句；第二个参数是传递给 SQL 语句的参数值，即需要寻找的用户 ID。

2. Mapper 实例

SqlSession 可以通过代理的方式来执行 SQL 语句，从而可以通过 Mapper 接口发送 SQL 语句。例如：

```
UserMapper userMapper = SqlSession.getMapper(UserMapper.class);
User user = userMapper.selectById (3);
```

在上述代码中，UserMapper 是一个接口，定义如下：

```
public interface UserMapper{
    User selectById(Integer userId);
}
```

在这种情况下，要求接口的包名与映射器的命名空间相同，接口的方法名与映射器中的 SQL ID 相同，接口的方法参数与 SQL 语句的参数相同，这样才能正确定位和执行 SQL 语句。这种接口被称为 Mapper 接口，所获得的实例被称为 Mapper 实例（实际上它是一个代理实例）。

在非 Spring 环境下，需要手动创建 SqlSessionFactory 和 SqlSession；在 Spring 环境下，只需要进行一定配置，就可以由容器创建 SqlSessionFactory 和 SqlSession。而且，容器能够创建 Mapper 实例，并将实例自动注入业务对象中。也就是说，在 Spring 环境下，只需要进行一定配置，再定义好 Mapper 接口和映射器，就可以实现数据库操作。

在 MyBatis3 之前，只支持 XML 映射器，即所有的 SQL 语句都必须在 XML 文件中配置。而从 MyBatis3 开始，还支持接口映射器，这种映射器允许以 Java 代码的方式定义 SQL 语句，即将 SQL 语句描述在 Mapper 接口中，非常简洁。例如：

```
public interface UserMapper{
    @Select("select * FROM test_user where user_id = #{userId}")
    User selectById(Integer userId);
}
```

在这里，Mapper 接口具有映射器的作用。

在 Spring Boot 中使用 MyBatis 会更加简洁，这是因为它省去了大量的配置工作。Spring Boot 会自动注册 SqlSessionFactoryBean、SqlSessionTemplate、ClassPathMapperScanner 等组件。SqlSessionFactoryBean 用于加载用户自定义的配置，构建 SqlSessionFactory。SqlSessionTemplate 是 MyBatis 的代理类，是 Spring 提供的一个对 MyBatis 的 SqlSession 的增强类。它的作用是将 SqlSession 与当前的事务绑定，而且是线程安全的。一个 SqlSessionTemplate 可以被多个 Dao 所共享。ClassPathMapperScanner 提供 MyBatis 的 Mapper 自动扫描功能，并且由 MapperScannerConfigurer 将每个 Mapper 包装成一个类型为 MapperFactoryBean 的 BeanDefinition，再注册到 IoC 容器中。在创建 Mapper 实例时，使用 MapperFactoryBean 获取了 Mapper 代理实例（当调用 MapperFactoryBean 的 getObject()方法时，MyBatis 会使用 JDK 的动态代理创建 Mapper 代理对象）。

4.2 Spring Boot 与 MyBatis 整合

4.2.1 MyBatis 基本配置

在 Spring Boot 中使用 MyBatis 时，相应的组件可以帮助我们完成很多配置工作，使得 MyBatis 使用起来更加简洁。

1．引入依赖

使用 MyBatis 需引入如下依赖：

```xml
<!-- MyBatis 依赖 -->
<dependency>
    <groupId>org.mybatis.spring.boot</groupId>
    <artifactId>mybatis-spring-boot-starter</artifactId>
    <version>2.1.1</version>
</dependency>
```

2．配置 MyBatis

Spring Boot 提供了 MyBatis 的自动配置类 MybatisAutoConfiguration，可以自动注册 SqlSessionFactoryBean、SqlSessionTemplate 等组件，开发人员只需在配置文件中指定相关属性即可。

数据源配置与 Spring Data JPA 的配置相同。MyBatis 的主要配置如下：

```
#实体类所在的包
mybatis.type-aliases-package=com.rc.entity
#开启 MyBatis 的驼峰命名转换
mybatis.configuration.map-underscore-to-camel-case=true
#配置实体与数据库表的 XML 映射文件的位置
mybatis.mapperLocations = classpath:mapper/*.xml
```

4.2.2 基于 MyBatis 实现 Dao 层

在 Spring Boot 中使用 MyBatis 实现 Dao 层主要有三项工作：定义实体类、编写映射器、编写 Mapper 接口。

1．定义实体类

在使用 MyBatis 时，实体类的定义不需要像使用 Spring Data JPA 一样添加映射标注，只需要属性的命名遵循小驼峰规则，并与表中的字段对应即可。遵循小驼峰规则，即单词首字母小写，后面的每个单词首字母大写。在与表中的字段对应时，字段的每个单词之间用下画线分隔。例如，用户实体类的代码如下：

```java
@Data
public class RcUser {
    private Integer userId;
    private String userName;
    private String userRealname;
    private String userPwd;
```

```
        private String userType;
        private String userEmail;
        private String userPosttime;
        private RcCompany company;
        @ToString.Exclude
        private List<RcRole> roles;
}
```

2. 编写映射器

映射器文件一般是 XML 文件（称为 Mapper 文件），主要用于编写 SQL 语句，一些结果集与实体类的映射关系，以及一些缓存配置等。SQL 语句的描述及映射关系也可以使用标注的方式编写在 Mapper 接口的方法之前。例如：

```xml
<?xml version="1.0" encoding="UTF-8"?>
<!DOCTYPE mapper PUBLIC "-//mybatis.org//DTD Mapper 3.0//EN"
        "http://mybatis.org/dtd/mybatis-3-mapper.dtd">
<mapper namespace="com.rc.dao.RcUserDao">
    <resultMap type="RcUser" id="userResultMap" autoMapping="false">
        <id property="userId" column="user_id" />
        <result property="userName" column="user_name"/>
        <result property="userPwd" column="user_pwd"/>
        <result property="userType" column="user_type"/>
        <result property="userEmail" column="user_email"/>
        <result property="userPosttime" column="user_posttime"/>
        <association property="person" column="per_id" javaType="com.rc.entity.RcCompany"
            select="com.rc.dao.RcCompanyDao.selectById">
        </association>
        <collection property="roles" column="user_id" javaType="ArrayList" ofType="RcRole"
            select="com.rc.dao.RcRoleDao.selectRoles">
        </collection>
    </resultMap>
    <insert id="insert" parameterType=" com.rc.entity.RcUser">
        insert into rc_user(user_name,user_pwd,user_type,user_email,user_posttime,com_id)
        values(#{userName},#{userPwd},#{userType},#{userEmail},#{userPosttime},
        #{company.comId})
    </insert>
    <update id="update" parameterType=" com.rc.entity.RcUser">
        update rc_user set user_name=#{userName},user_pwd=#{userPwd},user_type=#{userType},
        user_email=#{userEmail},com_id = #{company.comId}
        where user_id=#{userId}
    </update>
    <update id="updateRole">
        delete from rc_user_role where user_id=#{userId};
        insert into rc_user_role(user_id,role_id) values
        <foreach collection="roleIds" item="roleId" separator="," close=";">
            (#{userId},#{roleId})
        </foreach>
    </update>
    <delete id="delete">
```

```xml
        delete from rc_user where user_id = #{userId}
    </delete>
    <select id="selectById" resultMap="userResultMap" parameterType="java.lang.Integer">
        select * from rc_user where user_id=#{userId}
    </select>
    <select id="selectByUserName" resultMap="userResultMap" parameterType="java.lang.String" useCache="false">
        select * from rc_user where user_name=#{userName}
    </select>
    <select id="selectOne" resultMap="userResultMap" useCache="false">
        select * from rc_user where user_name=#{userName} and #{userPwd}
    </select>
    <select id="selectByPage" resultMap="userResultMap" useCache="false">
        select * from rc_user
        <where>
            <if test="userName!=null">
                user_name like CONCAT(CONCAT('%', #{userName}), '%')
            </if>
        </where>
        order by user_posttime desc
        <if test="startRecord!=null and pageSize!=null">
            limit #{startRecord},#{pageSize}
        </if>
    </select>
    <select id="selectCount" resultType="java.lang.Long" parameterType="hashmap">
        select count(*) from rc_user
        <where>
            <if test="userName!=null">
                user_name like CONCAT(CONCAT('%', #{userName}), '%')
            </if>
        </where>
    </select>
</mapper>
```

3．编写 Mapper 接口

Mapper 接口描述了业务可以使用的方法。MyBatis 根据接口定义创建接口的动态代理对象，Mapper 接口方法对应映射器文件中描述的 SQL 语句。

Mapper 接口开发需要遵循以下规范。

- Mapper 文件中的 namespace 与 Mapper 接口的类路径相同。
- Mapper 接口的方法名与 Mapper 文件中定义的每条 SQL 语句的 ID 相同。
- Mapper 接口方法的输入参数类型与 Mapper 文件中定义的每条 SQL 语句的 parameterType 的类型相同。
- Mapper 接口方法的输出参数类型和 Mapper 文件中定义的每条 SQL 语句的 resultType 的类型相同。

例如，管理员的 Mapper 接口定义如下：

```
@Mapper
```

```java
public interface RcUserDao {
    void insert(RcUser user);
    void update(RcUser user);
    void updateRole(Integer userId, Integer[] roleIds);
    void delete(Integer userId);
    RcUser selectById(Integer userId);
    RcUser selectByUserName(String userName);
    RcUser selectOne(@Param("userName") String username, @Param("userPwd") String userPwd);
    List<RcUser> selectSomeByPage(String userName, Integer pageNo, Integer pageSize);
    Integer selectCount(String userName);
}
```

在接口定义时使用@Mapper 标注是为了让容器能够找到 Mapper 接口。Mapper 接口一般被存放在根包（启动类所在的包）的子包下。否则，需要使用@MapperScan 标注指定要扫描的 Mapper 接口的路径。例如：

```java
@SpringBootApplication
@MapperScan("com.test.dao")
public class RcApplication {
    public static void main(String[] args) {
        SpringApplication.run(RcApplication.class, args);
    }
}
```

在 Spring Boot 中，定义了 Mapper 接口，就表示完成了 Dao 层的设计。也就是说，Mapper 接口就是 Dao 接口。有了接口后，就不需要再定义实现类了。

4.2.3 MyBatis 映射器配置

1. 基于 XML 文件的映射器

SQL 映射文件包含的主要元素如下所述。

- mapper：映射文件的根元素节点只有一个属性 namespace（命名空间），用于区分不同的 mapper，并且 namespace 全局唯一。在绑定接口时，namespace 的名称必须与接口相同。
- cache：配置指定命名空间的缓存。
- cache-ref：从其他命名空间中引用缓存配置。
- resultMap：用于描述数据库结果集和对象的对应关系。
- sql：可以重用的 SQL 块，也可以被其他语句引用。
- insert：映射插入语句。
- update：映射更新语句。
- delete：映射删除语句。
- select：映射查询语句。

1）select 元素

select 元素的主要属性如下所述。

- id：SQL 语句的标识，在命名空间中是唯一的，可以被用来引用这条语句。

- parameterType：参数类型完全限定名或别名。
- resultType：返回值类型。
- usecache：是否为二级缓存。

例如，使用 select 完成单条件模糊查询，代码如下：

```
<select id="getAdminListByUserName" resultType="userResult" parameterType="String">
    select * from rc_user where userName like CONCAT ('%',#{userName},'%')
<select>
```

上述代码定义了一个 id 为 getUserListByUserName 的查询语句，参数类型为 String，返回值类型是 userResult。

在 MyBatis 中可以使用${}和#{}两种方法获取参数：#{}基于 PreparedStatement 的占位符实现；${}基于字符串拼接实现。如果 parameterType 是一个对象，则#{}、${}都可以用来获取对象的属性。由于#{}是通过 PreparedStatement 实现的，可以有效防止 SQL 注入，而${}不可以，因此一般使用#{}。

select 元素的参数有以下 5 种情况。

- 单一的简单参数，如 Integer 类型、String 类型等。

使用#{参数名}取值即可，因为#{}只是占位符，所以{}里的名称与参数名是否相同没有关系。但如果在使用 if 判断时，则需要使用_parameter 取值判断。例如：

```
<select id="selectListByUserName" resultMap="personResult" parameterType="java.lang.String" >
    select * from rc_person where per_sex = '男'
    <if test="_parameter != null and _parameter != "" >
        and per_nature like CONCAT( '%', #{perNature, jdbcType=VARCHAR} , '%')
    </if>
    order by per_posttime desc
</select>
```

- 将查询条件封装成对象（parameterType="对象的类名"）。例如：parameterType="RcPerson"。

这时，取属性值格式如下：#{属性}。

- 使用 Map 集合存储查询条件（parameterType="Map"）。

这时，直接通过 Map 中的 key 取值，格式如下：#{key}。

- 带有多个参数，接口中没有使用@Param 标注参数。在这种情况下不能使用参数名，但可以使用#{param1},#{param2}…或者#{0},#{1}…取值。
- 使用@Param 标注实现多参数传递，接口的定义代码如下：

```
public int updatePwd(@Param("id")Integer id,@param("userPwd") String pwd);
```

这时，直接通过参数名取值，格式如下：#{id}和#{userPwd}。

2）resultMap 元素

resultMap 元素的主要属性如下所述。

- id：结果集映射的唯一标识。在一个映射器文件中可以定义多个结果集映射，每个映射的 ID 是唯一的。
- type：映射的实体对象类型。

其中，type 的主要属性如下所述。

- id：主键映射。
- result：一般字段映射。
- collection：一对多映射。
- association：多对一映射。

例如，下面是用户结果映射，在查询用户时可查出其拥有的角色：

```xml
<resultMap id="userResultMap" type="com.rc.user.entity.RcUser">
    <id column="user_id" property="userId"/>
    <result column="user_name" property="userName"/>
    <result column="user_realname" property="userRealname"/>
    <result column="user_pwd" property="userPwd"/>
    <result column="user_type" property="userType"/>
    <result column="user_email" property="userEmail"/>
    <result column="user_posttime" property="userPosttime"/>
    <result column="com_id" property="comId"/>
    <result column="com_name" property="comName"/>
    <collection property="roles" column="user_id" javaType="ArrayList"
                ofType="com.rc.user.entity.RcRole"
                select="com.rc.user.dao.RcRoleDao.selectRolesByUserId">
    </collection>
</resultMap>
```

其中，collection 的主要属性如下所述。

- property：一个对象集合属性。
- column：主键名。
- javaType：集合类型。
- ofType：关联的对象类名。
- select：SQL ID。

再如，下面是申请结果映射，在查询申请时可查出其对应的人和工作职位：

```xml
<resultMap type="com.rc.entity.RcApplication" id="appplicationResultMap" autoMapping="false">
    <id column="app_id" property="appId" />
    <result column="app_postTime" property="appPosttime" />
    <result column="app_response" property="appResponse" />
    <result column="app_rspTime" property="appRsptime" />
    <result column="app_post_time" property="appPostTime" />
    <result column="app_rsp_time" property="appRspTime" />
    <association property="person" column="per_id"
        javaType="com.rc.entity.RcPerson"
        select="com.rc.dao.RcPersonDao.selectById">
    </association>
    <association property="job" column="job_id"
        javaType="com.rc.entity.RcJob"
        select="com.rc.dao.RcJobDao.selectById">
    </association>
</resultMap>
```

其中，association 的主要属性如下所述。

- property：一个对象属性。

- column：当前表的主键，对应关联表的外键。
- javaType：关联的对象类型。
- select：SQL ID。

3）insert 元素

insert 元素的主要属性如下所述。

- id：SQL 语句的标识，在命名空间中是唯一的，可以被用来引用这条语句。
- parameterType：参数类型完全限定名或别名。
- flushCache：是否刷新缓存，默认值为 true。
- useGeneratedKeys：是否开启数据库内部生成的自增主键，默认值为 false。
- keyProperty：仅针对 insert 和 update，表示主键属性。keyProperty 表示将使用 POJO 的哪个属性去匹配这个主键，说明它会用数据库生成的主键去赋值给 POJO 的这个属性。

例如，下面的代码为插入用户：

```xml
<insert id="insert" parameterType=" com.rc.entity.RcUser">
    insert into rc_user(user_name,user_pwd,user_type,user_email,user_posttime,com_id)
    values(#{userName},#{userPwd},#{userType},#{userEmail},#{userPosttime},#{company.comId})
</insert>
```

如果在插入用户后想要获取主键，可以使用如下方式：

```xml
<insert id="insert" useGeneratedKeys="true" keyProperty="userId"
                    parameterType="com.rc.entity.RcUser">
    insert into rc_user(user_name,user_pwd,user_type,user_email,user_posttime,com_id)
    values(#{userName},#{userPwd},#{userType},#{userEmail},#{userPosttime},#{company.comId})
</insert>
```

4）update 和 delete 元素

这两个元素的主要属性如下所述。

- id：SQL 语句的标识，在命名空间中是唯一的，可以被用来引用这条语句。
- parameterType：参数类型完全限定名或别名。
- flushCache：是否刷新缓存，默认值为 true。

例如：

```xml
<!-- 修改用户 -->
<update id="update" parameterType="com.rc.entity.RcUser">
    update rc_user set user_name=#{userName},user_pwd=#{userPwd},user_type=#{userType},
    user_email=#{userEmail},com_id = #{company.comId}
    where user_id=#{userId}
</update>
<!-- 修改用户角色 -->
<update id="updateRole">
    delete from rc_user_role where user_id=#{userId};
    insert into rc_user_role(user_id,role_id) values
    <foreach collection="roleIds" item="roleId" separator="," close=";">
        (#{userId},#{roleId})
    </foreach>
</update>
```

```xml
<!-- 删除用户 -->
<delete id="delete">
    delete from rc_user where user_id = #{userId}
</delete>
```

其中，修改用户角色对应的接口方法如下：
void updateRole(@Param("userId") Integer userId, @Param("roleIds") Integer[] roleIds);

2. 基于标注的映射器

采用标注的方式可以将 SQL 映射编写在接口中。用到的主要标注如下所述。

- @Select：查询类的标注，所有的查询均使用这个标注。
- @Result：修饰返回的结果集，使实体类属性与数据库字段名一一对应。如果实体类属性和数据库字段名保持一致，就不需要使用这个属性来修饰。
- @Insert：负责插入。如果直接传入实体类，则会自动解析属性为对应的值。
- @Update：负责修改，也可以直接传入对象。
- @Delete：负责删除。

例如：

```java
@Mapper
public interface ContentDao {
    @Insert("insert into rc_admin(user_id, user_name, user_pwd, user_type) values(#{admin.userName}, #{admin.userPwd}, #{admin.userType})")
    void insert(RcAdmin admin);
    @Update("update rc_admin set user_id=#{admin.userName}, user_name=#{admin.userPwd}, user_type=#{admin.userType} where user_id=#{admin.userId}")
    void update(RcAdmin admin);
    @UpdateProvider(type=MySelectProvider.class,method="selectSome")
    @Insert({
            "<script>",
            "delete from rc_admin_role where user_id=#{userId};",
            "insert into rc_admin_role(user_id,role_id) values ",
            "<foreach collection='roleIds' item='id' index='index' separator=','>",
            "(#{userId}, #{id})",
            "</foreach>",
            "</script>"
    })
    void update(@Param("userId") Integer userId,@Param("roleIds") Integer[] roleIds);
    @Delete("delete from rc_admin where user_id=#{userId}")
    void delete(Integer userId);
    @Select("select * from rc_admin where user_id=#{userId}")
    @Results({
            @Result(id=true, column="user_id", property="userId"),
            @Result(column="user_name",property="userName"),
            @Result(column="user_pwd",property="userPwd"),
            @Result(column="user_type",property="userType"),
            @Result(column="user_id",property="roles",
                    many=@Many(
                            select="com.rc.dao.getRolesByUserId",
                            fetchType= FetchType.LAZY
```

```
            )
        )
})
RcAdmin selectById(Integer userId);
@Select("select * from rc_admin where user_name=#{userName} and user_pwd=#{userPwd}")
RcAdmin selectOne(@Param("userName") String userName, @Param("userPwd") String userPwd);
@Select("select * from rc_admin where limit #{recordStart},#{pageSize}")
List<RcAdmin> selectByPage(@Param("recordStart") Integer recordStart,
@Param("pageSize") Integer pageSize);
@Select("select count(*) from rc_admin")
Integer selectCount();
}
```

4.3　基于 MyBatis Generator 的逆向工程

4.3.1　MyBatis Generator 基础

1．MyBatis Generator 简介

MyBatis Generator 是 MyBatis 和 iBatis 的代码生成器。它将为所有版本的 MyBatis 及 2.2.0 版本之后的 iBatis 生成代码。它将内部缺省数据库表（或许多表），并生成可用于访问数据库表的组件。这减少了设置对象、配置文件，以及与数据库表交互的初始麻烦。

MyBatis Generator 主要完成的工作是依据数据库表创建对应的 Model、DAO 和映射文件，可以通过 Maven 插件或 mybatis-generator 的 jar 包生成。

MyBatis Generator 的使用较为简单，生成的 DAO 及映射文件中包含基本的 CRUD 操作。对于复杂的应用而言，需要在此基础上进行修改。需要注意的是，在一次项目中应当避免多次执行 mybatis-generator，即应当尽量在数据库表创建完整且确定不会修改后执行 mybatis-generator，否则再次执行会覆盖原本的 Model、DAO 和映射文件。

2．在 Spring Boot 项目中使用 MyBatis Generator

主要步骤如下所述。

（1）引入 MyBatis Generator 插件依赖，代码如下：

```
<dependency>
        <groupId>org.mybatis.generator</groupId>
        <artifactId>mybatis-generator-core</artifactId>
        <version>1.3.7</version>
</dependency>
```

（2）创建 mybatis-generator.xml 文件。

在配置文件中指定数据库连接地址、生成类的存放地址、生成对应表的类名等信息，代码如下：

```
<?xml version="1.0" encoding="UTF-8"?>
<!DOCTYPE generatorConfiguration
        PUBLIC "-//mybatis.org//DTD MyBatis Generator Configuration 1.0//EN"
        "http://mybatis.org/dtd/mybatis-generator-config_1_0.dtd">
```

```xml
<generatorConfiguration>
    <!-- 一个数据库一个 context -->
    <!-- flat: 所有内容（主键，blob）等全部生成在一个对象中 -->
    <context id="sqlserverTables" defaultModelType="flat">
        <!-- 生成的 Java 文件的编码 -->
        <property name="javaFileEncoding" value="UTF-8"/>
        <!-- 格式化 Java 代码 -->
        <property name="javaFormatter" value="org.mybatis.generator.api.dom.DefaultJavaFormatter"/>
        <!-- 格式化 XML 代码 -->
        <property name="xmlFormatter" value="org.mybatis.generator.api.dom.DefaultXmlFormatter"/>
        <!-- beginningDelimiter 和 endingDelimiter: 指明数据库中用于标记数据库对象名的符号，比如 Oracle 默认是双引号，MySQL 默认是反引号 -->
        <property name="beginningDelimiter" value="`"/>
        <property name="endingDelimiter" value="`"/>
        <!--生成 mapper.xml 文件时覆盖原文件-->
        <plugin type="org.mybatis.generator.plugins.UnmergeableXmlMappersPlugin" />
        <plugin type="nnn.page.MapperAnnotationPlugin"/>
        <!--optional，指在创建 class 时，对注释进行控制-->
        <commentGenerator>
            <property name="suppressDate" value="true"/>
            <!-- 是否去除自动生成的注释 true：是；false：否 -->
            <property name="suppressAllComments" value="true"/>
        </commentGenerator>
        <!-- 数据库连接 -->
        <jdbcConnection driverClass="com.mysql.cj.jdbc.Driver"
            connectionURL="jdbc:mysql://localhost:3306/rc?useUnicode=true&amp;characterEncoding=utf8&amp;serverTimezone=GMT&amp;useSSL=false&amp;zeroDateTimeBehavior=convertToNull&amp;allowMultiQueries=true"
            userId="root"
            password="12345">
        </jdbcConnection>
        <!-- 类型转换 -->
        <javaTypeResolver>
            <!--
                true：使用 BigDecimal 对应 DECIMAL 和 NUMERIC 数据类型；
                false：默认；
                scale>0;length>18: 使用 BigDecimal;
                scale=0;length[10,18]: 使用 Long;
                scale=0;length[5,9]: 使用 Integer;
                scale=0;length<5: 使用 Short;
            -->
            <property name="forceBigDecimals" value="false" />
        </javaTypeResolver>
        <!--模型生成器，targetPackage 指定包，targetProject 指定路径 -->
        <javaModelGenerator targetPackage="com.rc.entity"
                            targetProject="./rc/src/main/java">
            <!--true：在 targetPackage 的基础上，根据数据库的 schema 再生成一层 package，最终生成的类存放在这个 package 下，默认为 false -->
            <property name="enableSubPackages" value="false" />
```

```xml
        <!-- 从数据库返回的值被清理前后的空格 -->
        <property name="trimStrings" value="true" />
        <!-- 是否生成构造函数 默认是false -->
        <property name="constructorBased" value="false"/>
    </javaModelGenerator>
    <!-- 生成映射文件（XML 文件），注意，在 MyBatis3 之后，可以不使用 mapper.xml
文件，若使用 mapper.xml 文件，这个元素就必须配置。targetProject 同 javaModelGenerator-->
    <sqlMapGenerator targetPackage="mapper" targetProject=".rc/src/main/resources">
        <property name="enableSubPackages" value="false" />
        <property name="constructorBased" value="false"/>
</sqlMapGenerator>
<!-- Mapper 接口类文件生成器
    targetPackage：同 javaModelGenerator；
    type：选择怎么生成 Mapper 接口（在 MyBatis3/MyBatis3Simple 下）；
    1）ANNOTATEDMAPPER：会生成使用 Mapper 接口+Annotation 的方式创建
（SQL 生成在 annotation 中），不会生成对应的 XML
    2）MIXEDMAPPER：使用混合配置，会生成 Mapper 接口，并适当添加合适的
Annotation，但是 XML 会生成在 XML 中
    3）XMLMAPPER：会生成 Mapper 接口，接口完全依赖 XML 文件
-->
<javaClientGenerator type="XMLMAPPER"
        targetPackage="com.rc.dao" targetProject="./rc/ src/main/java">
    <property name="enableSubPackages" value="false" />
    <property name="constructorBased" value="false"/>
</javaClientGenerator>
<!-- 选择一个 table 来生成相关文件，可以有一个或多个 table，但必须有 table 元素。
    tableName（必要）：生成对象的表名；
    domainObjectName：生成的 Domain 类的名称；
    enableInsert（默认为 true）：指定是否生成 insert 语句；
    enableSelectByPrimaryKey（默认为 true）：指定是否生成按照主键查询对象的
语句（即 getById 或 get）；
    enableSelectByExample（默认为 true）：MyBatis3Simple 为 false, 指定是否生成
动态查询语句；
    enableUpdateByPrimaryKey（默认为 true）：指定是否生成按照主键修改对象的
语句（即 update);
    enableDeleteByPrimaryKey（默认为 true）：指定是否生成按照主键删除对象的
语句（即 delete）；
    enableDeleteByExample（默认为 true）：MyBatis3Simple 为 false, 指定是否生
成动态删除语句；
    enableCountByExample（默认为 true）：MyBatis3Simple 为 false, 指定是否生成
动态查询总条数语句（用于分页的总条数查询）；
    enableUpdateByExample（默认 true）：MyBatis3Simple 为 false, 指定是否生成
动态修改语句（只修改对象中不为空的属性）；
-->
<table tableName="yslnt_category"  domainObjectName="YslntCategory"
    enableCountByExample="false" enableUpdateByExample="false"
    enableDeleteByExample="false" enableSelectByExample="true"
    selectByExampleQueryId="false" mapperName="yslntCategoryMapper">
```

```
                <!-- 相当于在生成的 insert 元素上添加 useGeneratedKeys="true"和
        keyProperty 属性-->
                <!-- <generatedKey column="cont_id" sqlStatement="MySql" identity=
        "true"/>-->
            </table>
            …
        </context>
    </generatorConfiguration>
```

（3）运行。运行方式包括以下两种。

一种是通过插件运行。首先在 pom 文件中配置插件，然后打开 Maven 视窗，在找到插件后，单击鼠标右键并在弹出的快捷菜单中选择相应命令来运行，如图 4-2 所示。

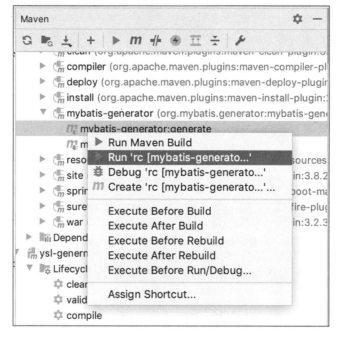

图 4-2　通过插件运行

例如：

```
<plugin>
    <groupId>org.mybatis.generator</groupId>
    <artifactId>mybatis-generator-maven-plugin</artifactId>
    <version>1.3.5</version>
    <configuration>
        <!-- 在控制台打印执行日志 -->
        <verbose>true</verbose>
        <!-- 在重复生成时会覆盖之前的文件-->
        <overwrite>true</overwrite>
        <configurationFile>src/main/resources/mybatis-generator.xml
    </configurationFile>
    </configuration>
    <!-- 数据库连接选择 8.0 版本以上的，因为使用的是 mysql8.0 -->
```

```xml
        <dependencies>
            <dependency>
                <groupId>mysql</groupId>
                <artifactId>mysql-connector-java</artifactId>
                <version>8.0.16</version>
            </dependency>
        </dependencies>
</plugin>
```

另一种是通过实例运行。这需要建立一个类，并在该类中包含 main()方法，在 main()方法中通过建立对象来运行。例如：

```java
public class GeneratorApplication {
    public void generator() throws Exception {
        List<String> warnings = new ArrayList<String>();
        boolean overwrite = true;
        InputStream in = this.getClass().getResourceAsStream("mybatis-generator.xml");
        System.out.println(in == null);
        ConfigurationParser cp = new ConfigurationParser(warnings);
        Configuration config = cp.parseConfiguration(in);
        DefaultShellCallback callback = new DefaultShellCallback(overwrite);
        MyBatisGenerator1 myBatisGenerator = new MyBatisGenerator1(config, callback, warnings);
        myBatisGenerator.generate(null);
    }
    public static void main(String[] args) throws Exception {
        try {
            GeneratorApplication generatorApplication = new GeneratorApplication();
            generatorApplication.generator();
        } catch (Exception e) {
            e.printStackTrace();
        }
    }
}
```

4.3.2 MyBatis Generator 扩展

1. 调整数据访问层接口名称格式

默认生成的数据访问层接口名称为 XxxxMapper 格式。若需要改为 XxxxDao 格式，则可以自定义一个插件，用来继承 PluginAdapter。插件代码如下：

```java
public class RenameJavaMapperPlugin extends PluginAdapter {
    private String searchString;
    private String replaceString;
    private Pattern pattern;
    public RenameJavaMapperPlugin() {
    }
    public boolean validate(List<String> warnings) {
        searchString = properties.getProperty("searchString");
        replaceString = properties.getProperty("replaceString");
        boolean valid = stringHasValue(searchString)
```

```
                    && stringHasValue(replaceString);
            if (valid) {
                pattern = Pattern.compile(searchString);
            } else {
                if (!stringHasValue(searchString)) {
                    warnings.add(getString("ValidationError.18",
                            "RenameExampleClassPlugin",
                            "searchString"));
                }
                if (!stringHasValue(replaceString)) {
                    warnings.add(getString("ValidationError.18",
                            "RenameExampleClassPlugin",
                            "replaceString"));
                }
            }
            return valid;
        }
        @Override
        public void initialized(IntrospectedTable introspectedTable) {
            String oldType = introspectedTable.getMyBatis3JavaMapperType();
            Matcher matcher = pattern.matcher(oldType);
            oldType = matcher.replaceAll(replaceString);
            introspectedTable.setMyBatis3JavaMapperType(oldType);
        }
}
```

2. 在数据访问层接口类前添加@Mapper标注

为了自动在数据访问层接口类前添加@Mapper标注，可以自定义一个插件，用来继承PluginAdapter。插件代码如下：

```
public class MapperAnnotationPlugin extends PluginAdapter {
    @Override
    public boolean validate(List<String> warnings) {
        return true;
    }
    @Override
    public boolean clientGenerated(Interface interfaze, TopLevelClass topLevelClass, IntrospectedTable introspectedTable) {
        if (introspectedTable.getTargetRuntime() == TargetRuntime.MYBATIS3){
            interfaze.addImportedType(
                new FullyQualifiedJavaType("org.apache.ibatis.annotations.Mapper"));

            interfaze.addAnnotation("@Mapper");
        }
        return true;
    }
}
```

3. 解决分页问题

MyBatis支持使用RowBounds进行分页。MyBatis Generator可以通过插件机制来扩展这一功能。在 MyBatis Generator 的配置文件 generatorConfig.xml 中添加这个插件，就可以支持分页，代码如下：

```xml
<plugin type="org.mybatis.generator.plugins.RowBoundsPlugin"></plugin>
```

RowBounds 分页实质上是通过 ResultSet 的游标来实现分页的，也就是说，它并不是使用 select 语句的 limit 分页，而是使用 Java 代码分页。查询语句的结果集会包含符合查询条件的所有数据，若使用不当会导致性能问题，所以并不推荐使用 RowBoundsPlugin 来实现分页。

在实现 MySQL 分页时，更推荐使用 select 语句的 limit 来实现分页，然而 MyBatis Generator 目前并没有提供这样的插件。好在 MyBatis Generator 支持插件扩展，可以自己实现一个 limit 来定义分页的插件。这里给出另一种简便的方式，即使用 PageHelper 插件。

PageHelper 是一个 MyBatis 的分页插件，使用时很方便。PageHelper 通过拦截器获取同一线程中的预编译的 SQL 语句，然后将 SQL 语句包装成具有分页功能的 SQL 语句，并将其再次赋值给下一步操作，所以实际执行的 SQL 语句就是具有分页功能的 SQL 语句。

（1）引入依赖，代码如下：

```xml
<!--引入 pageHelper 分页插件 -->
<dependency>
    <groupId>com.github.pagehelper</groupId>
    <artifactId>pagehelper</artifactId>
    <version>5.1.11</version>
</dependency>
```

（2）在 application.properties 文件中进行配置，代码如下：

```
#配置分页插件
pagehelper.helperDialect=mysql
pagehelper.reasonable=true
pagehelper.supportMethodsArguments=true
pagehelper.params.count=countSql
```

（3）在业务层使用 PageHelper，代码如下：

```java
PageHelper.startPage(pageNo,pageSize);
RcCompanyExample c=new RcCompanyExample();
c.createCriteria().andComTypeLike("%"+comType+"%").andComNameLike("%"+comName+"%");
List<RcCompany> list=companyDao.selectByExample(c);
PageInfo pageInfo = new PageInfo(list);
```

查询得到的是 PageInfo 对象。该对象的主要属性如下所述。

- pageNum：当前页。
- pageSize：每页的数量。
- total：总记录数。
- pages：总页数。
- list：返回的数据 List<T>。

4.3.3 使用自动生成的代码操作数据库

MyBatis Generator 在使用数据库表生成数据代码时，除了生成实体的 POJO，还会同时生成 Example 类，以及在 mapper.xml 文件中生成 Example 类的 SQL 语句。

Example 类包含一个内部静态类 Criteria。使用 Criteria 类可以在 Example 类中根据我们自己的需求动态生成 SQL where 子句，不需要我们自己再修改 Mapper 文件以添加或者修改 SQL 语句了，从而节省很多编写 SQL 语句的时间。

下面介绍几种常用的方法。

（1）模糊搜索用户名，例如：

```
String name = "明";
RcPersonExample ex = new RcPersonExample();
ex.createCriteria().andUserRealnameLike('%'+name+'%');
List<RcPerson> list = personDao.selectByExample(ex);
```

（2）通过某个字段排序，例如：

```
String orderByClause = "per_posttime desc";
RcPersonExample ex = new RcPersonExample();
ex.setOrderByClause(orderByClause);
List<RcPerson> list = personDao.selectByExample(ex);
```

（3）条件搜索，不确定条件的个数，例如：

```
RcCompanyExample ex = new RcCompanyExample ();
Criteria criteria = ex.createCriteria();
if(StringUtils.isNotBlank(comType)){
    criteria.andComTypeLike("%"+comType+"%").
}
if(StringUtils.isNotBlank(comName)){
    criteria .andComNameLike("%"+comName+"%"
}
List<RcCompany> list = companyDao.selectByExample(ex);
```

4.4　基于 MyBatis-Plus 的逆向工程

4.4.1　MyBatis–Plus 基础

1. MyBatis-Plus 简介

MyBatis-Plus 是 MyBatis 的增强工具，在 MyBatis 的基础上增加了很多功能，并且简化了开发过程，提高了效率。MyBatis-Plus 将通用的 CRUD 操作封装进 BaseMapper 接口中，并且自动生成的 Mapper 接口自动继承了 BaseMapper 接口。复杂的 SQL 操作可以使用 QueryWrapper(LambdaQueryWrapper)和 UpdateWrapper(LambdaUpdateWrapper)方法进行动态 SQL 语句拼接。

MyBatis-Plus 架构如图 4-3 所示。

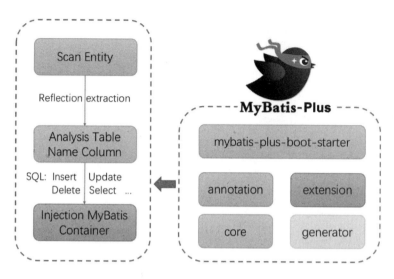

图 4-3　MyBatis-Plus 架构

MyBatis-Plus 的特点如下所述。

- 无侵入：MyBatis-Plus 只做增强不做改变，引入它不会对现有工程产生影响。
- 损耗小：启动 MyBatis-Plus，即可自动注入基本的 CURD 操作，性能基本无损耗，可以直接面向对象操作。
- 强大的 CRUD 操作：MyBatis-Plus 内置通用 Mapper 及 Service，仅通过少量配置即可实现单表的大部分 CRUD 操作，并且具有强大的条件构造器，可以满足各类使用需求。
- 支持 Lambda 形式调用：通过 Lambda 表达式可以方便地编写各类查询条件，无须担心写错字段。
- 支持主键自动生成：MyBatis-Plus 支持多达 4 种主键策略（内含分布式唯一 ID 生成器——Sequence），可以自由配置，更好地解决主键问题。
- 支持 ActiveRecord 模式：MyBatis-Plus 支持 ActiveRecord 形式调用，实体类只需要继承 Model 类即可进行强大的 CRUD 操作。
- 支持自定义全局通用操作：MyBatis-Plus 支持全局通用方法注入（Write once，use anywhere）。
- 内置代码生成器：MyBatis-Plus 使用代码或 Maven 插件可以快速生成 Mapper 层、Model 层、Service 层和 Controller 层代码，支持模板引擎，包含很多自定义配置。
- 内置分页插件：MyBatis-Plus 基于 MyBatis 物理分页，使得开发者无须关心具体操作。在配置好插件后，编写分页等同于普通的 List 查询。
- 支持多种数据库：MyBatis-Plus 支持 MySQL、MariaDB、Oracle、DB2、H2、HSQL、SQLite、Postgre、SQLServer 等多种数据库。
- 内置性能分析插件：MyBatis-Plus 可输出 SQL 语句及其执行时间，建议在开发测试时启用该功能，能够快速找出慢查询。

- 内置全局拦截插件：MyBatis-Plus 提供全表 Delete、Update 操作的智能分析阻断功能，也可以自定义拦截规则，预防误操作。

2. 在 Spring Boot 项目中使用 MyBatis-Plus

（1）引入依赖，代码如下：

```xml
<dependency>
    <groupId>com.baomidou</groupId>
    <artifactId>mybatis-plus-boot-starter</artifactId>
    <version>3.3.1</version>
</dependency>
<!-- 这是 MyBatis-Plus 的代码自动生成器 -->
<dependency>
    <groupId>com.baomidou</groupId>
    <artifactId>mybatis-plus-generator</artifactId>
    <version>3.3.1</version>
</dependency>
```

（2）在 application.properties 文件中进行配置，代码如下：

```
#扫描位置
mybatis-plus.mapper-locations=classpath:/mapper/*Mapper.xml
mybatis-plus.typeAliasesPackage=com.rc.entity
#数据库类型
mybatis-plus.global-config.db-config.db-type=MYSQL
#逻辑删除指定值
mybatis-plus.global-config.db-config.logic-delete-value=1
mybatis-plus.global-config.db-config.logic-not-delete-value=0
#下画线转驼峰
mybatis-plus.configuration.map-underscore-to-camel-case=true
```

（3）创建一个配置类，代码如下：

```java
public class MyGenerator {
    public static void main(String[] args) {
        //数据库、用户名、密码，生成代码存放位置和包名
        gen("rc","root","12345","./src/main", "com.rc");
    }
    private static void gen(String db,String username,String password,String outputdir, String basePackageName) {
        String classPath = outputdir + File.separator + "java" + File.separator + String.join
(File.separator, basePackageName.split("\\."));
        System.out.println(classPath);
        //1.全局配置
        GlobalConfig gconfig = new GlobalConfig();
        gconfig.setActiveRecord(true)              //是否支持 AR 模式
                .setAuthor("yang")                  //设置作者
                .setFileOverride(true)              //是否覆盖文件
                .setActiveRecord(true)              //开启 activeRecord 模式
                .setOutputDir(outputdir)            //生成路径
                .setEnableCache(false)              //XML 二级缓存
                .setIdType(IdType.AUTO)             //主键策略
                .setMapperName("%sDao")             //持久层接口名称格式
```

```
                .setServiceName("%sService")           //Service 接口名格式
                .setServiceImplName("%sServiceImpl")   //Service 实现名格式
                .setControllerName("%sController")     //Controller 名称格式
                .setXmlName("%sMapper")                //映射文件名格式
                .setSwagger2(true)                     //整合 Swagger2
                .setBaseResultMap(true)                //生成基本的结果集
                .setBaseColumnList(true);              //生成基本的列的集合 SQL 片段
//2.数据源配置
DataSourceConfig dsconfig = new DataSourceConfig();
dsconfig.setDbType(DbType.MYSQL)
        .setDriverName("com.mysql.jdbc.Driver")
        .setUsername(username)                         //用户名
        .setPassword(password)                         //密码
        .setUrl("jdbc:mysql://localhost:3306/"+db+"?useUnicode=true&characterEncoding=utf-8&useSSL=false");
//3.包名策略配置
PackageConfig pkconfig = new PackageConfig();
pkconfig.setParent(basePackageName)                    //设置父包，后面就不用写这段包名了
        .setMapper("dao")
        .setService("service")
        .setServiceImpl("service.impl")
        .setController("controller")
        .setEntity("entity")
        .setXml("mapper");


//4.设置不同类文件生成的路径
HashMap<String, String> pathMap = new HashMap<>();
pathMap.put(ConstVal.ENTITY_PATH, classPath + "/entity");
pathMap.put(ConstVal.MAPPER_PATH, classPath + "/dao");
pathMap.put(ConstVal.SERVICE_PATH, classPath + "/service");
pathMap.put(ConstVal.SERVICE_IMPL_PATH, classPath + "/service/impl");
pathMap.put(ConstVal.CONTROLLER_PATH, classPath + "/controller");
pathMap.put(ConstVal.XML_PATH, outputdir + "/resources/mapper");
pkconfig.setPathInfo(pathMap);
//5.自定义需要填充的字段
List<TableFill> tableFillList = new ArrayList<TableFill>();
TableFill createField = new TableFill("gmt_create", FieldFill.INSERT);
TableFill modifiedField = new TableFill("gmt_modified", FieldFill.INSERT_UPDATE);
tableFillList.add(createField);
tableFillList.add(modifiedField);
//6.策略配置
StrategyConfig strategy = new StrategyConfig();
strategy.setNaming(NamingStrategy.underline_to_camel)//驼峰命名
        .setColumnNaming(NamingStrategy.underline_to_camel)
        //实体类父类
        //.setSuperEntityClass("com.rc.entity.RcBaseEntity")
        //写于父类中的公共字段
        //.setSuperEntityColumns("user_id")
        .setEntityLombokModel(true)
```

```
                    .setRestControllerStyle(true)
                    .setTableFillList(tableFillList)
                    //公共父类
                    .setSuperControllerClass("com.rc.controller.RcBaseController")
                    .setSuperServiceClass(null)
                    .setSuperServiceImplClass(null)
                    .setSuperMapperClass(null)
                    .setControllerMappingHyphenStyle(true)
                    .setInclude("user_info".split(","));
        //7.配置模板
        TemplateConfig templateConfig = new TemplateConfig();
        templateConfig.setController("/mytemplates/controller.vm");
        ag.setTemplate(templateConfig);
        ag.execute();
        //8.注入自定义参数
        InjectionConfig injectionConfig = new InjectionConfig() {
            //自定义属性注入: abc
            //在.ftl（或者.vm）模板中，通过${cfg.abc}获取属性
            @Override
            public void initMap() {
                Map<String, Object> map = new HashMap<>();
                map.put("restControllerStyle", false);
                this.setMap(map);
            }
        };
        //9.整合配置
        AutoGenerator ag = new AutoGenerator();
        ag.setGlobalConfig(gconfig)
            .setDataSource(dsconfig)
            .setPackageInfo(pkconfig)
            .setStrategy(strategy);
    }
}
```

（4）运行 MyBatis-Plus。

在通过 MyBatis-Plus 代码生成器生成代码时，可以参考官方文档。生成的代码主要包含 entity、mapper、service 和 controller。这样基本的代码就有了，可以根据需要进行修改。

4.4.2 MyBatis–Plus 扩展

在通过 MyBatis-Plus 代码生成器生成代码时，entity、dao 和 service 能够满足要求，但 controller 只是一个框架，我们可以自定义模板。MyBatis-Plus 支持 Velocity（默认）、FreeMarker 和 Beetl 三种模板引擎，如果这 3 种模板引擎都不满足要求，还可以采用自定义模板引擎。这里我们使用 Velocity 模板引擎，需引入如下依赖：

```
<dependency>
    <groupId>org.apache.velocity</groupId>
    <artifactId>velocity-engine-core</artifactId>
```

```
        <version>2.2</version>
    </dependency>
```

MyBatis-Plus 默认的模板被存储在 mybatis-plus-generator/templates 中,我们可以将该文件夹中的模板复制出来,存储在 java/main/resources/mytemplates 中,并在此基础上进行一些修改,也可以完全自定义。

下面的代码是针对控制层制定的模板,文件名称为 controller.vm:

```
package ${package.Controller};
import com.baomidou.mybatisplus.core.metadata.IPage;
import com.baomidou.mybatisplus.extension.plugins.pagination.Page;
import org.springframework.ui.ModelMap;
import com.rc.util.PageList;
import org.springframework.web.servlet.ModelAndView;
import ${package.Service}.${table.serviceName};
import ${package.Entity}.${entity};
import io.swagger.annotations.Api;
import io.swagger.annotations.ApiOperation;
import lombok.extern.slf4j.Slf4j;
import org.springframework.beans.factory.annotation.Autowired;
import org.springframework.web.bind.annotation.*;
#if(${superControllerClassPackage})
import ${superControllerClassPackage};
#end
import java.util.List;
import java.util.ArrayList;
import java.util.Arrays;
/**
 * 名称: ${table.controllerName}
 * 描述: ${table.comment}控制类
 *
 * @version 1.0
 * @author: ${author}
 * @datetime: ${date}
 */
@Slf4j
@Api(tags = {"${table.comment}"})
#if(${restControllerStyle})
@RestController
#else
@Controller
#end
@RequestMapping("/${table.entityPath}")
#if(${superControllerClass})
public class ${table.controllerName} extends ${superControllerClass} {
#else
public class ${table.controllerName} {
#end
    @Autowired
    public ${table.serviceName} ${table.entityPath}Service;
```

```java
@ApiOperation(value = "要添加")
@PostMapping("/willAdd")
public String willAdd(){
    return "redirect:/${table.name}/add";
}
@ApiOperation(value = "添加")
@PostMapping("/save")
public String save(${entity} ${table.entityPath}){
    ${table.entityPath}Service.save(${table.entityPath});
    return "redirect:/${table.entityPath}/manage";
}
@ApiOperation(value = "要修改")
@PostMapping("/willEdit/{id}")
public ModelAndView willEdt(@PathVariable("id") Integer id){
    ${entity} ${table.entityPath} = ${table.entityPath}Service.getById(id);
    return new ModelAndView("${table.name}/edit","${table.entityPath}",${table.entityPath});
}
@ApiOperation(value = "修改")
@PostMapping("/edit")
public String edit(${entity} ${table.entityPath}){
    ${table.entityPath}Service.updateById(${table.entityPath});
    return "redirect:/${table.entityPath}/manage";
}
@ApiOperation(value = "根据ID删除")
@PostMapping("/delete/{id}")
public String delete(@PathVariable("id") Integer id){
    ${table.entityPath}Service.removeById(id);
    return "redirect:/${table.entityPath}/manage";
}
@ApiOperation(value = "批量删除")
@PostMapping("/delete/batch")
public String deleteBatch(String ids){
    String[] idArray=ids.split(",");
    List<Integer> idList = new ArrayList<Integer>();
    for(String s:idArray){
        idList.add(Integer.valueOf(s));
    }
    ${table.entityPath}Service.removeByIds(idList);
    return "redirect:/${table.entityPath}/manage";
}
    @ApiOperation(value = "分页浏览")
    @GetMapping("/browse")
    public ModelAndView browse(Integer pageNo, Integer pageSize){
        ModelMap map = new ModelMap();
        IPage<${entity}> page = ${table.entityPath}Service.page(new Page<>(pageNo, pageSize));
        PageList<${entity}> plist = new PageList<${entity}>(page.getRecords(), page.getTotal(), pageNo, pageSize);
        map.put("pageNo", pageNo);
        map.put("pageSize", pageSize);
```

```
            map.put("plist", plist);
            return new ModelAndView("${table.name}/browse","map",map);
    }

    @ApiOperation(value = "管理")
    @GetMapping("/manage")
    public ModelAndView manage(Integer pageNo, Integer pageSize){
        ModelMap map = new ModelMap();
        IPage<${entity}> page = ${table.entityPath}Service.page(new Page<>(pageNo, pageSize));
        PageList<${entity}> plist = new PageList<${entity}>(page.getRecords(), page.getTotal(), pageNo, pageSize);
        map.put("pageNo", pageNo);
        map.put("pageSize", pageSize);
        map.put("plist", plist);
        return new ModelAndView("${table.name}/manage","map",map);
    }
    @ApiOperation(value = "按 ID 查询")
    @GetMapping("/findById/{id}")
    public ModelAndView findById(@PathVariable("id") Integer id){
        ${entity} ${table.entityPath} = ${table.entityPath}Service.getById(id);
        return new ModelAndView("${table.name}/show","${table.entityPath}",${table.entityPath});
    }
}
```

可按如下方式使用模板：

```
//注入自定义参数
InjectionConfig injectionConfig = new InjectionConfig() {
    //自定义属性注入：abc
    //在.ftl（或者是.vm）模板中，通过${cfg.abc}获取属性
    @Override
    public void initMap() {
        Map<String, Object> map = new HashMap<>();
        map.put("restControllerStyle", false);
        this.setMap(map);
    }
};
//配置模板
TemplateConfig templateConfig = new TemplateConfig();
templateConfig.setController("/mytemplates/controller.vm");
ag.setCfg(injectionConfig);    // ag 为 AutoGenerator 对象
ag.setTemplate(templateConfig);
```

4.4.3 基于 MyBatis-Plus 的数据操作

MyBatis-Plus 代码生成器已经帮我们封装好 Dao 层和 Service 层，我们只需要调用相应的方法即可。这里主要讨论一下复杂情况。

1. 分页处理

MyBatis-Plus 支持分页，首先需要进行配置，代码如下：

```java
@EnableTransactionManagement
@Configuration
public class MybatisPlusConfig {
    /**
     * mybatis-plus SQL 执行解释插件【生产环境可以关闭】
     */
    @Bean
    public SqlExplainInterceptor sqlExplainInterceptor() {
        return new SqlExplainInterceptor();
    }
    /**
     * 分页插件
     */
    @Bean
    public PaginationInterceptor paginationInterceptor() {
        return new PaginationInterceptor();
    }
}
```

然后需要在 Controller 层中编写方法，代码如下：

```java
@Controller
@RequestMapping("/admin")
public class RcAdminController {
    @Resource
    private RcAdminService adminService;
    @RequestMapping(value = "/manage", method = RequestMethod.POST)
    public ModelAndView manage(Integer pageNo,Integer pageSize){
        if(pageNo==null)pageNo=1;
        if(pageSize==null)pageSize=10;
        IPage<RcAdmin> page = new Page<>(pageNo, pageSize);
        map.put("page ", page);
        return new ModelAndView("admin/manage", "page", page);
    }
}
```

2. 执行复制查询

当查询条件复杂时，我们可以使用 MyBatis-Plus 的条件构造器，可以参考下面的 QueryWrapper 条件的参数说明：

```java
//查询性别为男的人
QueryWrapper<RcPerson> queryWrapper1 = new QueryWrapper<>();
queryWrapper1.eq("perSex", "男");
List<RcPerson> list1 = personService.list(queryWrapper1);
//查询在 1989 年—2020 年出生的人
QueryWrapper<RcPerson> queryWrapper2 = new QueryWrapper<>();
queryWrapper2.gt("perBirthdate", "1989-01-01");
queryWrapper2.le("perBirthdate", "2020-12-31");
List<RcPerson> list2 = personService.list(queryWrapper2);
```

```
//模糊查询姓张的人
QueryWrapper<RcPerson> queryWrapper3 = new QueryWrapper<>();
queryWrapper3.like("userName","张%");
queryWrapper3.orderByDesc("perPosttime");
List<RcPerson> userInfoEntityList3 = personService.list(queryWrapper3);
```

3. 自定义 SQL

引入 MyBatis-Plus 不仅不会对项目现有的 MyBatis 架构产生任何影响，而且 MyBatis-Plus 支持所有 MyBatis 原生的特性，这也是它受欢迎的原因之一。对于某些复杂业务而言，可以使用一些比较复杂的 SQL 语句，但是需要在 Dao 接口中添加方法的定义，然后在 Mapper 文件中添加 SQL 描述，或者以标注的方式直接添加到方法中。

第 5 章　基于 Spring Security 实现认证和授权

在 Web 应用开发中，安全无疑是十分重要的。使用 Spring Security 保护 Web 应用是一个非常好的选择。Spring Security 是 Spring 项目中的一个安全模块，可以非常方便地与 Spring 项目无缝集成。特别是在 Spring Boot 项目中加入 Spring Security 十分简单。本章将介绍 Spring Security 的基本原理，并结合网上人才中心系统介绍如何实现认证和授权。

5.1　Spring Security 概述

5.1.1　Spring Security 简介

Spring Security 是一种基于 Spring AOP 和 Servlet 过滤器的安全框架。它提供了全面的安全性解决方案，同时在 Web 请求级和方法调用级处理认证和授权。

Spring Security 提供了一组可以在 Spring 应用上下文中配置的 Bean，充分利用了 Spring IoC（Inversion of Control，控制反转）、DI（Dependency Injection，依赖注入）和 AOP（面向切面编程）功能，为应用系统提供了声明式的安全访问控制功能，减少了为企业系统的安全控制编写大量重复代码的工作。除了常规的认证和授权，Spring Security 还提供了很多高级特性（如 ACLs、LDAP、JAAS、CAS 等）以满足各种复杂场景下的安全需求。

Spring Security 是 Spring Boot 官方推荐使用的安全框架，具有以下特性。
- 支持与全面且可扩展的身份认证和授权支持。
- 防御会话固定、点击劫持、跨站请求伪造等攻击。
- 支持与 Servlet API 集成。
- 支持与 Spring Web MVC 集成。

5.1.2　Spring Security 原理

1．认证与授权

Spring Security 应用级别的安全性包括认证（Authentication）和授权（Authorization）两部分。认证指的是验证某个用户是否为系统中的合法主体，也就是说，用户能否访问该系统。认证一般要求用户提供用户名和密码，并由系统通过校验用户名和密码来完成认证过程。授权指的是验证某个用户是否具有权限执行某个操作。在一个系统中，不同用户所具有的权限是不同的。例如，对于一个文件来说，有的用户只能对其进行读取，而有的用户可以对其进行修改。一般来说，系统会为不同的用户分配不同的角色，而每个角色会对应一系列的权限。

Spring Security 的核心功能为认证和授权，所有的架构也是基于这两个核心功能来实现的。

授权策略主要基于角色的 RBAC（Role-Based Access Control，访问控制）。在 RBAC 中，权限与角色相关联，用户可以通过成为适当角色的成员而得到这些角色的权限，从而极大地简化了权限的管理。这样的权限管理是层级相互依赖的，先将权限赋予角色，再将角色赋予用户，使得权限设计更加清楚，管理起来更加方便。

2．Spring Security 过滤器

Spring Security 是通过过滤器来完成认证和授权的。Spring Security 在 Servlet 的过滤器链（Filter Chain）中注册了一个过滤器链代理 FilterChainProxy，它会为每次请求添加一条过滤器链，每次请求都需要通过过滤器链进行身份的认证和授权。不同的过滤器具有不同的功能，如下所述。

- WebAsyncManagerIntegrationFilter：将 Spring Security 上下文与 Spring Web 中用于处理异步请求映射的 WebAsyncManager 进行集成。
- SecurityContextPersistenceFilter：在每次请求处理之前，将与该请求相关的安全上下文信息加载到 SecurityContextHolder 中，然后在该次请求处理完成之后，将 SecurityContextHolder 中关于这次请求的信息存储到一个"仓储"中，最后将 SecurityContextHolder 中的信息清除，例如，使用该过滤器在 session 中维护一个用户的安全信息。
- HeaderWriterFilter：用于将头信息加入响应中。
- CsrfFilter：用于处理跨站请求伪造。
- LogoutFilter：用于处理退出登录。
- UsernamePasswordAuthenticationFilter：用于处理基于表单的登录请求，并从表单中获取用户名和密码。在默认情况下可以处理来自/login 的请求。在从表单中获取用户名和密码时，默认使用的表单 name 值为 username 和 password，这两个值可以通过设置该过滤器的 usernameParameter 和 passwordParameter 两个参数的值进行修改。
- DefaultLoginPageGeneratingFilter：如果没有配置登录页面，则在系统初始化时就会配置这个过滤器，用于在需要进行登录时生成一个登录表单页面。
- BasicAuthenticationFilter：用于检测和处理 HTTP Basic 认证。
- RequestCacheAwareFilter：用于处理请求的缓存。
- SecurityContextHolderAwareRequestFilter：主要用于包装请求对象 request。
- AnonymousAuthenticationFilter：用于检测 SecurityContextHolder 中是否存在 Authentication 对象，如果不存在，则为其提供一个匿名 Authentication。
- SessionManagementFilter：用于管理 session 的过滤器。
- ExceptionTranslationFilter：用于处理 AccessDeniedException 和 AuthenticationException 异常。
- FilterSecurityInterceptor：可以将其看作过滤器链的出口。

- RememberMeAuthenticationFilter：当用户没有登录而直接访问资源时，从 cookie 中找出用户的信息。如果 Spring Security 能够识别用户提供的 remember me cookie，则用户无须填写用户名和密码，就可以直接登录系统。该过滤器默认不开启。

3．认证与授权的主要过程

Spring Security 认证与授权的过程如图 5-1 所示。

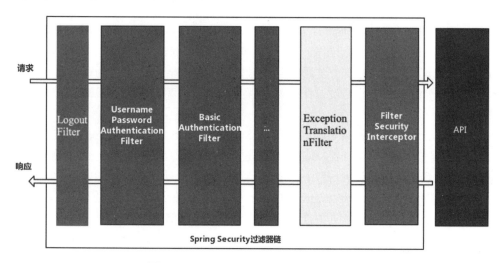

图 5-1　Spring Security 认证与授权的过程

客户端发起一个请求，进入 Spring Security 过滤器链。

（1）LogoutFilter 过滤器用于判断是否为注销路径，如果是注销路径，则交由 LogoutHandler 处理，否则直接进入下一个过滤器。如果注销成功，则交由 LogoutSuccessHandler 处理，否则转到 ExceptionTranslationFilter。

（2）UsernamePasswordAuthenticationFilter 过滤器用于判断是否为登录路径，如果是登录路径，则进入该过滤器进行登录操作，否则直接进入下一个过滤器。如果登录失败，则交由 AuthenticationFailureHandler 处理，否则交由 AuthenticationSuccessHandler 处理。

（3）FilterSecurityInterceptor 过滤器用于根据 URI 寻找对应的授权管理器，授权管理器会进行授权工作，如果授权成功则转到 Controller 层，否则交由 AccessDeniedHandler 处理。

5.1.3　Spring Security 配置基础

1．基本配置

引入 Spring Security 相关依赖，代码如下：

```
<dependency>
    <groupId>org.springframework.boot</groupId>
    <artifactId>spring-boot-starter-security</artifactId>
</dependency>
```

引入 jar 包，在 Spring Boot 项目启动后，对 Spring Security 进行自动配置。这时的应用程序已经加入 Spring Security 保护。当请求应用程序时，就会弹出一个登录界面，如图 5-2 所示。默认的用户名为 user，密码为在 Spring Security 启动时默认生成的密码（在启动日志中可以看到）。

图 5-2　登录界面

在 application.properties 文件中配置用户名和密码，代码如下：

```
#配置用户名
spring.security.user.name=admin
#配置密码
spring.security.user.password=123
#配置角色
spring.security.user.roles=admin
```

如果希望在配置自己的登录界面时配置多个用户，并针对不同的用户赋予不同的权限，就需要自定义一个类用来继承 WebSecurityConfigurerAdapter，并通过覆盖以下 3 个方法来完成更多的配置。

- configure(AuthenticationManagerBuilder auth)：配置用户。
- configure(HttpSecurity http)：配置访问控制。
- configure(WebSecurity web)：配置静态资源的控制方式。

2．认证配置

用户在登录时，会被登录验证拦截器 AuthenticationProcessingFilter 拦截，然后进入认证过程。Spring Security 将认证方式定义为实现了 AuthenticationProvider 接口的 Provider。管理这些认证方式的是 ProviderManager，它实现了 AuthenticationManager 接口。

ProviderManager 内部存放着一个 Provider 数组，验证的过程为遍历 Provider，并调用其 authenticate() 方法。当一个合适的 Provider 验证通过后，就会将用户的权限信息封装为一个 Authentication 并存放到 Spring 的全局缓存 SecurityContextHolder 中，以备后面访问资源时使用。

Spring Security 提供了一个 AuthenticationManagerBuilder 类，用来构建 Authentication Manager。Spring Security 提供了多种认证方式（如内存、JDBC、LDAP 和自定义 UserDetailsService 验证），但不管是哪一种验证方式，都是通过一个自动注入的 AuthenticationManagerBuilder

对象来完成的。在 Spring Boot 中自定义一个类，用来继承 WebSecurityConfigurerAdapter，并覆盖 configure（AuthenticationManagerBuilder auth）方法，然后利用这个方法的 AuthenticationManagerBuilder 参数 auth，就可以很方便地对认证进行配置。

认证的配置主要是对用户名、密码及权限的配置，有 3 种配置方法。

（1）直接配置在内存中。例如：

```java
@EnableWebSecurity
public class SecurityConfiguration extends WebSecurityConfigurerAdapter {
    @Override
    protected void configure(AuthenticationManagerBuilder auth) throws Exception {
        auth.inMemoryAuthentication()
            .withUser("wang").password("{noop}123").roles("admin")
            .and()
            .withUser("zhang").password("{noop}456").roles("user");
    }
}
```

上面的代码在内存中配置了两个用户，密码采用的是明文。

在 Spring Security 中提供了 BCryptPasswordEncoder 密码编码工具，如果采用加密的方式，则可以采用如下代码：

```java
@EnableWebSecurity
public class SecurityConfiguration extends WebSecurityConfigurerAdapter {
    @Bean
    public PasswordEncoder passwordEncoder(){
        return new BCryptPasswordEncoder();
    }
    @Override
    protected void configure(AuthenticationManagerBuilder auth) throws Exception {
        auth.inMemoryAuthentication()
            .withUser("wang").password("$2a$10$5gF9kaZ1reXBZhY8zQ2eu.af/oQStFOM5FcObw4wu7QPSAEGLUBRO").roles("admin")
            .and()
            .withUser("zhang").password("$2a$10$5gF9kaZ1reXBZhY8zQ2eu.af/oQStFOM5FcObw4wu7QPSAEGLUBRO").roles("user");
    }
}
```

（2）直接指定数据源和查询语句。用户信息及权限存放在数据库中，可以采用如下代码：

```
auth.jdbcAuthentication().dataSource(dataSource)
    .usersByUsernameQuery(query1).authoritiesByUsernameQuery(query2);
```

其中，dataSource 是数据源，query1 为根据用户名查询用户的 SQL 语句，query2 为根据用户名查询用户权限的 SQL 语句。例如：

```java
@EnableWebSecurity
public class SecurityConfiguration extends WebSecurityConfigurerAdapter {
    ...
    @Override
    protected void configure(AuthenticationManagerBuilder auth) throws Exception {
        auth.jdbcAuthentication().dataSource(dataSource)
            .usersByUsernameQuery("select username,password,enabled from users WHERE username=?")
```

```
        .authoritiesByUsernameQuery("select username,authority from authorities where username=?")
        .passwordEncoder(passwordEncoder());
    }
}
```

（3）定义一个 UserService 类，实现 UserDetailsService 接口。这个接口有一个方法，格式如下：

UserDetails loadUserByUsername(String username) throws UsernameNotFoundException

这个方法用于根据用户名查找用户，并且该用户类实现了 UserDetials 接口。例如：

```
@Service
public class UserService implements UserDetailsService{
    ...
    @Override
    public UserDetails loadUserByUsername(String username) throws UsernameNotFoundException{
        ...
    }
}
```

上面的 UserService 类添加了@Service 标注，因此当应用启动时，容器就会将 UserService 对象注入 AuthenticationManager。当然，我们也可以明确指定。例如：

```
@EnableWebSecurity
public class SecurityConfiguration extends WebSecurityConfigurerAdapter {
    private UserDetailsService userService;
    ...
    @Override
    protected void configure(AuthenticationManagerBuilder auth) throws Exception {
        auth.userDetailsService(userService).passwordEncoder(passwordEncoder());
    }
}
```

如果使用这种方式进行配置，则 Spring Security 会启用 DaoAuthenticationProvider 认证器。该认证器首先调用 UserDetailsService.loadUserByUsername()方法查询用户，然后使用 PasswordEncoder.matches()方法进行密码比对。如果认证成功，则返回一个 Authentication 对象（认证过的且含有权限信息），AuthenticationManager 会将其存放到 Spring 的全局缓存 SecurityContextHolder 中，以备后面访问资源时使用。

3．授权配置

当登录成功并访问资源（即授权管理）时，会被拦截器 AbstractSecurityInterceptor 拦截，并且会调用 FilterInvocationSecurityMetadataSource 的方法来获取被拦截的 URL 所需的全部权限，然后调用授权管理器 AccessDecisionManager。这个授权管理器不仅会被通过 Spring 的全局缓存 SecurityContextHolder 获取用户的权限信息，还会获取被拦截的 URL 和被拦截 URL 所需的全部权限，并根据设置的投票策略 AccessDecisionVoter 进行授权决策。如果权限足够，则返回；如果权限不够，则报错并调用权限不足页面。

Spring Security 提供了 HttpSecurity，可以方便地配置资源的访问权限。在继承 WebSecurityConfigurerAdapter 类的类中覆盖 configure(HttpSecurity http)方法，并使用该方法的参数 http 就可以完成基本的配置。这个方法不仅可以对授权规则进行配置，还可以对登录、退出及过滤器等进行配置。例如：

```
@Override
protected void configure(HttpSecurity http) throws Exception {
    http
        .formLogin()
            .loginPage("/login_page")
            .passwordParameter("username")
            .passwordParameter("password")
            .loginProcessingUrl("/sign_in")
            .permitAll()
            .and()
        .authorizeRequests()
            .antMatchers("/test").hasRole("test")
            .anyRequest().authenticated()
            .and()
        .logout()
            .logoutSuccessUrl("/")
            .and()
        .csrf().disable();
    http.addFilterAt(getAuthenticationFilter(),UsernamePasswordAuthenticationFilter.class);
    http.exceptionHandling().accessDeniedHandler(new MyAccessDeniedHandler());
    http.addFilterAfter(new MyFittler(), LogoutFilter.class);
}
```

在上面的配置中，首先对登录进行了配置，包括登录页请求路径、用户名属性名、密码属性名及登录请求路径等，permitAll()方法表示任意用户可以访问。然后配置了资源权限，包括/test url 应该具有什么权限才能访问；anyRequest()方法、authenticated()方法表示已登录的用户才能访问其他资源。接下来配置了退出后访问的路径。最后添加了两个过滤器。

如果希望在数据库中配置资源权限，则需要完成以下工作。

（1）自定义 FilterSecurityInterceptor，继承 AbstractSecurityInterceptor 类并实现 Filter 接口。其主要目的是将自定义的 SecurityMetadataSource 与自定义的 accessDecisionManager 配置到自定义的拦截器 FilterSecurityInterceptor 中。

（2）自定义 SecurityMetadataSource，实现 FilterInvocationSecurityMetadataSource 接口。其目的是从数据库或其他数据源中加载资源权限。

（3）自定义 MyAccessDecisionManager，实现 AccessDecisionManager 接口。在一般情况下，不需要定义 AccessDecisionVoter，只需要在 decide()方法中直接编写授权决策代码即可。decide()方法没有返回值，在未通过授权时抛出 AccessDeniedException。如果访问某个资源需要同时拥有两个或两个以上的权限，就需要自定义 AccessDecisionVoter 来实现。

4．静态资源控制配置

WebSecurityConfigurerAdapter 提供了 3 个 configure()方法，分别用来对 AuthenticationManagerBuilder、WebSecurity 与 HttpSecurity 进行配置。

其中，AuthenticationManagerBuilder 是账号认证配置，WebSecurity 主要是与 Web 应用相关的配置，HttpSecurity 则是对所有 HTTP 请求进行管理的相关配置。HttpSecurity 的

ignoring()方法与 WebSecurity 的 permitAll()方法都可以控制静态资源。两者的本质区别如下所述。

WebSecurity 的 ignoring()方法完全绕过了 Spring Security 的所有 Filter，相当于不经过认证，比较适合配置与前端相关的静态资源。

HttpSecurity 的 permitAll()方法则没有绕过 Spring Security，其中包含登录的及匿名的请求。该方法会给没有登录的用户配置一个 AnonymousAuthenticationToken，并将其设置到 SecurityContextHolder，方便后面的 Filter 统一处理 Authentication。

因此，静态资源控制配置一般使用 WebSecurity 的 ignoring()方法。例如：

```
@Override
//在这里配置哪些页面不需要认证
public void configure(WebSecurity web) throws Exception {
    //解决静态资源被拦截的问题
    web.ignoring().antMatchers("/**/*.js", "/**/*.css", "/**/*.jpg", "/**/*.png");
}
```

5.2 网上人才中心系统权限体系设计与开发

本节以网上人才中心系统为例来进行说明。数据层还是使用 MyBatis-Plus。

5.2.1 权限相关数据结构及实体类设计

1. 权限相关数据结构设计

与权限相关的数据库表主要有用户表、角色表、权限表、用户角色中间表和角色权限中间表。这些表的 EER 图如图 5-3 所示，数据存取情况如图 5-4、图 5-5、图 5-6、图 5-7、图 5-8 所示。

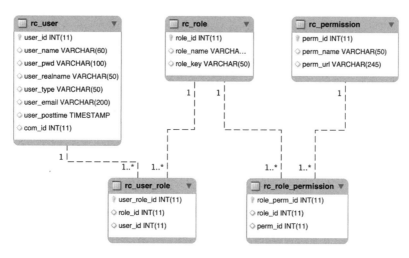

图 5-3　与权限相关的数据库表的 EER 图

user_id	user_name	user_pwd	user_realname	user_type	user_email	user_posttime	com_id
1	yang	$2a$10$.CAsj.EYmbd4sMd/pCPn9egFGRpC...	yang	系统用户	yang@ss.com	2020-03-27 08:10:22	7
2	li	$2a$10$.CAsj.EYmbd4sMd/pCPn9egFGRpC...	li	企业用户	li@ss.com	2020-03-27 08:10:41	7
3	zhang	$2a$10$.CAsj.EYmbd4sMd/pCPn9egFGRpC...	zhang	企业用户	zhang@ss.com	2020-04-13 15:42:37	7
4	zhao	$2a$10$.CAsj.EYmbd4sMd/pCPn9egFGRpC...	zhao	企业用户	zhao@ss.com	2020-04-13 15:42:44	8
5	wang	$2a$10$.CAsj.EYmbd4sMd/pCPn9egFGRpC...	wang	企业用户	wang@ss.com	2020-04-13 15:43:08	5

图 5-4　用户表的数据存取情况

role_id	role_name	role_key
1	管理员	ROLE_ADMIN
2	个人用户	ROLE_USER
3	企业用户	ROLE_COMPANY
4	超级管理员	ROLE_SUPERADMIN

图 5-5　角色表的数据存取情况

perm_id	perm_name	perm_url
1	系统管理	/sysmanage
2	用户管理	/rcUser/manage
3	企业信息管理	/rcCompany/manage
4	新闻管理	/rcNews/manage
6	角色管理	/rcRole/manage
7	权限管理	/rcPermission/manage
8	企业管理	/companymanage
9	招聘管理	/rcJob/manage
10	应聘管理	/rcApplication/manage
11	个人中心	/center
12	个人修改	/rcUser/personalWillEdit
13	完善信息	/rcPerson/willSetInfo
14	我的申请	/rcApplication/browse
15	修改密码	/rcUser/willEditPwd
16	浏览企业	/rcCompany/browse
17	浏览招聘	/rcJob/browse
19	浏览人才	/rcPerson/browse
20	浏览新闻	/rcNews/browse

图 5-6　权限表的数据存取情况

user_role_id	role_id	user_id
1	1	2
2	2	2
3	2	3
4	3	3
5	2	4
6	3	4
7	2	5
8	3	5
9	1	1
10	2	1
11	4	1

图 5-7　用户角色中间表的数据存取情况

第 5 章 基于 Spring Security 实现认证和授权

	role_perm_id	role_id	perm_id
1	1	1	1
2	2	1	4
3	3	1	4
4	4	2	11
5	5	2	12
6	6	2	13
7	7	2	14
8	8	2	15
9	9	2	16
10	10	2	17
11	11	2	20
12	12	3	8
13	13	3	9
14	14	3	10
15	15	3	19
16	16	3	3
17	17	4	2
18	18	4	4
19	19	4	4
20	20	4	6
21	21	4	7

图 5-8　角色权限中间表的数据存取情况

2．权限相关实体类设计

这里主要涉及用户、角色和权限 3 个实体类。此外，还有一个用户详细类，该类实现了 UserDetails 接口。权限相关实体类如图 5-9 所示。

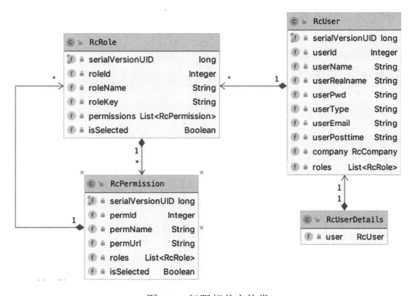

图 5-9　权限相关实体类

1）用户实体类及用户详细类

用户实体类代码如下：

```
@Data
@EqualsAndHashCode(callSuper = false)
```

```java
@Accessors(chain = true)
@TableName(value = "rc_user",resultMap = "BaseResultMap")
public class RcUser extends Model<RcUser>{
    @TableId(value = "user_id", type = IdType.AUTO)
    private Integer userId;
    private String userName;
    private String userRealname;
    private String userPwd;
    private String userType;
    private String userEmail;
    private String userPosttime;
    @ToString.Exclude
    private RcCompany company;
    @ToString.Exclude
    private List<RcRole> roles;
    @Override
    protected Serializable pkVal() {
        return this.getUserId();
    }
}
```

用户详细类用于 Spring Security 登录认证，代码如下：

```java
public class RcUserDetails implements UserDetails {
    private RcUser user;
    public RcUserDetails(RcUser user) {
        this.user = user;
    }
    @Override
    public Collection<? extends GrantedAuthority> getAuthorities() {
        List<SimpleGrantedAuthority> authorities = new ArrayList<>();
        for (RcRole role : user.getRoles()) {
            authorities.add(new SimpleGrantedAuthority(role.getRoleKey()));
        }
        return authorities;
    }
    @Override
    public String getPassword() {
        return user.getUserPwd();
    }
    @Override
    public String getUsername() {
        return user.getUserName();
    }
    @Override
    public boolean isAccountNonExpired() {
        return true;
    }
    @Override
    public boolean isAccountNonLocked() {
        return true;
```

```java
    }
    @Override
    public boolean isCredentialsNonExpired() {
        return true;
    }
    @Override
    public boolean isEnabled() {
        return true;
    }
    public RcUser getUser() {
        return user;
    }
}
```

2）角色实体类

角色实体类的代码如下：

```java
@Data
@EqualsAndHashCode(callSuper = false)
@Accessors(chain = true)
@TableName(value = "rc_role",resultMap = "BaseResultMap")
public class RcRole extends Model<RcRole> {
    private static final long serialVersionUID=1L;
    @ApiModelProperty(value = "角色 ID")
    @TableId(value = "role_id", type = IdType.AUTO)
    private Integer roleId;
    @ApiModelProperty(value = "角色名称")
    private String roleName;
    @ApiModelProperty(value = "角色 Key")
    private String roleKey;
    @ToString.Exclude
    private List<RcPermission> permissions;
    private Boolean isSelected;
    @Override
    protected Serializable pkVal() {
        return this.roleId;
    }
}
```

3）权限实体类

权限实体类的代码如下：

```java
@Data
@EqualsAndHashCode(callSuper = false)
@Accessors(chain = true)
public class RcPermission extends Model<RcPermission> {
    @TableId(value = "perm_id", type = IdType.AUTO)
    private Integer permId;
    private String permName;
    private String permUrl;
    private List<RcRole> roles;
    @Override
```

```
protected Serializable pkVal() {
    return this.permId;
}
}
```

5.2.2 权限相关数据访问层设计

1．映射器设计

1）用户映射器文件

用户映射器文件主要涉及用户与角色一对多关联、修改用户角色，以及根据用户名查询用户，代码如下：

```xml
<?xml version="1.0" encoding="UTF-8"?>
<!DOCTYPE mapper PUBLIC "-//mybatis.org//DTD Mapper 3.0//EN" "http://mybatis.org/dtd/mybatis-3-mapper.dtd">
<mapper namespace="com.rc.dao.RcUserDao">
    <!-- 通用查询映射结果 -->
    <resultMap id="BaseResultMap" type="com.rc.entity.RcUser">
        <id column="user_id" property="userId"/>
        <result column="user_name" property="userName"/>
        <result column="user_realname" property="userRealname"/>
        <result column="user_pwd" property="userPwd"/>
        <result column="user_type" property="userType"/>
        <result column="user_email" property="userEmail"/>
        <result column="user_posttime" property="userPosttime"/>
        <association property="company" column="com_id" javaType="com.rc.entity.RcCompany"
                    select="com.rc.dao.RcCompanyDao.selectById">
        </association>
        <collection property="roles" column="user_id" javaType="ArrayList"
                   ofType="com.rc.entity.RcRole" select="com.rc.dao.RcRoleDao.selectRolesByUserId">
        </collection>
    </resultMap>
    <resultMap id="BaseResultMap1" type="com.rc.entity.RcUser">
        <id column="user_id" property="userId"/>
        <result column="user_name" property="userName"/>
        <result column="user_realname" property="userRealname"/>
        <result column="user_pwd" property="userPwd"/>
        <result column="user_type" property="userType"/>
        <result column="user_email" property="userEmail"/>
        <result column="user_posttime" property="userPosttime"/>
    </resultMap>
    <!-- 通用查询结果列 -->
    <sql id="Base_Column_List">
        user_id, user_name, user_pwd, user_realname, user_type, user_email, user_posttime
    </sql>
    <update id="updateRole">
        delete from rc_user_role where user_id=#{userId};
        insert into rc_user_role(user_id,role_id) values
```

```xml
            <foreach collection="roleIds" item="roleId" separator="," close=";">
                (#{userId},#{roleId})
            </foreach>
        </update>
        <insert id="insertUser" useGeneratedKeys="true" keyProperty="userId">
            insert into rc_user(user_name, user_pwd, user_realname, user_type, user_email, com_id)
            values (#{userName}, #{userPwd}, #{userRealname}, #{userType},
             #{userEmail}, #{company.comId})
        </insert>
        <insert id="updateUser">
            update rc_user set user_name=#{userName}, user_pwd=#{userPwd},
            user_realname=#{userRealname}, user_type=#{userType}, user_email=#{userEmail},
                com_id=#{company.comId} where user_id = #{userId}
        </insert>
        <select id="seletByUserName" parameterType="String" resultMap="BaseResultMap">
            select * from rc_user where user_name = #{userName}
        </select>
        <!-- 用于企业与用户关联 -->
        <select id="selectUsersByComId" parameterType="Integer" resultMap="BaseResultMap">
            select * from rc_user where com_id = ${comId}
        </select>
        <!-- 用于个人用户（人才）与用户关联 -->
        <select id="selectByUserId" parameterType="Integer" resultMap="BaseResultMap1">
            select * from rc_user where user_id = ${userId}
        </select>
</mapper>
```

2）角色映射器文件

角色映射器文件主要涉及角色与权限多对多映射、修改权限、根据用户 ID 查询角色，以及根据权限 ID 查询角色，代码如下：

```xml
<?xml version="1.0" encoding="UTF-8"?>
<!DOCTYPE mapper PUBLIC "-//mybatis.org//DTD Mapper 3.0//EN" "http://mybatis.org/dtd/mybatis-3-mapper.dtd">
<mapper namespace="com.rc.dao.RcRoleDao">
    <!-- 通用查询映射结果 -->
    <resultMap id="BaseResultMap" type="com.rc.entity.RcRole">
        <id column="role_id" property="roleId" />
        <result column="role_name" property="roleName" />
        <result column="role_key" property="roleKey" />
        <collection property="permissions" column="role_id" javaType="ArrayList"
            ofType="com.rc.entity.RcPermission"
            select="com.rc.dao.RcPermissionDao.selectPermissionsByRoleId">
        </collection>
    </resultMap>
    <resultMap id="BaseResultMap1" type="com.rc.entity.RcRole">
        <id column="role_id" property="roleId" />
        <result column="role_name" property="roleName" />
        <result column="role_key" property="roleKey" />
    </resultMap>
```

```xml
<!-- 通用查询结果列 -->
<sql id="Base_Column_List">
    role_id, role_name, role_key
</sql>
<!-- 修改角色拥有的权限 -->
<update id="updatePermission">
    delete from rc_role_permission where role_id=#{roleId};
    insert into rc_role_permission(role_id,perm_id) values
    <foreach collection="permIds" item="permId" separator="," close=";">
        (#{roleId},#{permId})
    </foreach>
</update>
<!-- 用于用户与角色关联 -->
<select id="selectRolesByUserId" parameterType="Integer" resultMap="BaseResultMap">
    select b.* from rc_user_role a,rc_role b where b.role_id=a.role_id and a.user_id=#{userId}
</select>
<!-- 用于权限与角色关联 -->
<select id="selectRolesByPermId" parameterType="Integer" resultMap="BaseResultMap1">
    select b.* from rc_role_permission a,rc_role b where b.role_id=a.role_id and a.perm_id=#{permId}
</select>
</mapper>
```

3）权限映射器文件

权限映射器文件主要涉及权限与角色一对多关联、按角色查询权限，代码如下：

```xml
<?xml version="1.0" encoding="UTF-8"?>
<!DOCTYPE mapper PUBLIC "-//mybatis.org//DTD Mapper 3.0//EN" "http://mybatis.org/dtd/mybatis-3-mapper.dtd">
<mapper namespace="com.rc.dao.RcPermissionDao">
    <!-- 通用查询映射结果 -->
    <resultMap id="BaseResultMap" type="com.rc.entity.RcPermission">
        <id column="perm_id" property="permId" />
        <result column="perm_name" property="permName" />
        <result column="perm_url" property="permUrl" />
        <collection property="roles" column="perm_id" javaType="ArrayList" ofType="com.rc.entity.RcRole"
            select="com.rc.dao.RcRoleDao.selectRolesByPermId">
        </collection>
    </resultMap>
    <resultMap id="BaseResultMap1" type="com.rc.entity.RcPermission">
        <id column="perm_id" property="permId" />
        <result column="perm_name" property="permName" />
        <result column="perm_url" property="permUrl" />
    </resultMap>
    <!--用于角色与权限关联-->
    <select id="selectPermissionsByRoleId" parameterType="Integer" resultMap="BaseResultMap1">
        select b.* from rc_role_permission a,rc_permission b where b.perm_id=a.perm_id and a.role_id=#{roleId}
    </select>
    <!-- 通用查询结果列 -->
    <sql id="Base_Column_List">
        perm_id, perm_name, perm_url
```

```
        </sql>
</mapper>
```

2. 权限相关数据访问层接口设计

权限相关数据访问层接口的继承关系如图 5-10 所示，权限相关数据访问层接口如图 5-11 所示。

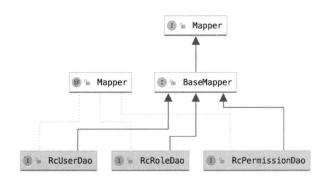

图 5-10　权限相关数据访问层接口的继承关系

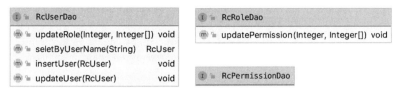

图 5-11　权限相关数据访问层接口

1）用户数据访问接口

用户数据访问接口的代码如下：

```
@Mapper
public interface RcUserDao extends BaseMapper<RcUser> {
    void updateRole(@Param("userId") Integer userId, @Param("roleIds") Integer[] roleIds);
    RcUser seletByUserName(String userName);
    void insertUser(RcUser user);
    void updateUser(RcUser user);
}
```

2）角色数据访问接口

角色数据访问接口的代码如下：

```
@Mapper
public interface RcRoleDao extends BaseMapper<RcRole> {
    void updatePermission(@Param("roleId")Integer roleId, @Param("permIds") Integer[] permIds);
}
```

3）权限数据访问接口

权限数据访问接口的代码如下：

```
@Mapper
public interface RcPermissionDao extends BaseMapper<RcPermission> {
}
```

5.2.3 权限相关业务逻辑层设计

权限相关业务逻辑层接口如图 5-12 所示。

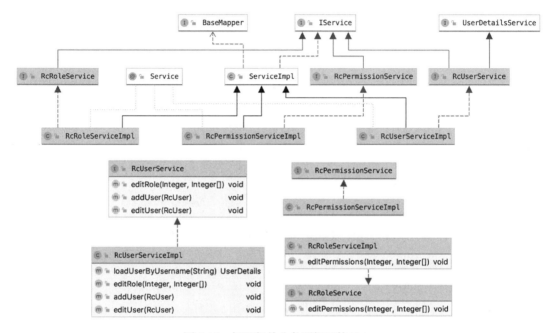

图 5-12 权限相关业务逻辑层接口

1．业务逻辑层接口设计

1）用户业务逻辑接口

用户业务逻辑接口 RcUserService 继承于 UserDetailsService 接口，代码如下：

```
public interface RcUserService extends IService<RcUser>,UserDetailsService{
    void editRole(Integer userId,Integer[] roleIds);
    void addUser(RcUser user);
    void editUser(RcUser user);
}
```

2）角色业务逻辑接口

角色业务逻辑接口的代码如下：

```
public interface RcRoleService extends IService<RcRole> {
    void editPermissions(Integer roleId,Integer[] permIds);
}
```

3）权限业务逻辑接口

权限业务逻辑接口的代码如下：

```
public interface RcPermissionService extends IService<RcPermission> {
}
```

2. 业务逻辑层实现类设计

1) 用户业务逻辑实现类

用户业务逻辑实现类可以实现 RcUserService 接口，其中，loadUserByUsername(String username)方法是 UserDetailsService 接口的方法，用于登录认证，代码如下：

```java
@Service
public class RcUserServiceImpl extends ServiceImpl<RcUserDao, RcUser> implements RcUserService {
    @Override
    public UserDetails loadUserByUsername(String userName) throws UsernameNotFoundException {
        RcUser user = this.baseMapper.seletByUserName(userName);
        if (user == null) {
            throw new UsernameNotFoundException("账号不存在");
        }
        return new RcUserDetails(user);
    }
    @Override
    public void editRole(Integer userId, Integer[] roleIds) {
        this.baseMapper.updateRole(userId, roleIds);
    }

    @Override
    public void addUser(RcUser user) {
        this.baseMapper.insertUser(user);
    }

    @Override
    public void editUser(RcUser user) {
        this.baseMapper.updateUser(user);
    }
}
```

2) 角色业务逻辑实现类

角色业务逻辑实现类代码如下：

```java
@Service
public class RcRoleServiceImpl extends ServiceImpl<RcRoleDao, RcRole> implements RcRoleService {
    @Override
    public void editPermissions(Integer roleId, Integer[] permIds) {
        this.baseMapper.updatePermission(roleId,permIds);
    }
}
```

3) 权限业务逻辑实现类

权限业务逻辑实现类代码如下：

```java
@Service
public class RcPermissionServiceImpl extends ServiceImpl<RcPermissionDao, RcPermission> implements RcPermissionService {
}
```

5.2.4 权限相关控制层设计

1. 用户控制类

用户控制类的代码如下:

```java
@Slf4j
@Api(tags = {"用户控制类"})
@Controller
@RequestMapping("/rcUser")
public class RcUserController extends RcBaseController {
    @Resource
    private RcUserService rcUserService;
    @Resource
    private RcRoleService rcRoleService;
    @Resource
    private RcCompanyService rcCompanyService;
    public PasswordEncoder passwordEncoder = new BCryptPasswordEncoder();
    @ApiOperation(value = "要注册")
    @GetMapping("/willRegister")
    public String willRegister() {
        return "rc_user/register";
    }
    @ApiOperation(value = "注册")
    @PostMapping("/register")
    @Transactional
    public ModelAndView register(RcUser rcUser) {
        rcUser.setUserType("个人用户");
        rcUser.setUserPwd(passwordEncoder.encode(rcUser.getUserPwd()));
        rcUserService.addUser(rcUser);
        return new ModelAndView("common/message", "msg", "注册成功! <a href='/rc'>返回首页</a>");
    }
    @ApiOperation(value = "要添加")
    @GetMapping("/willAdd")
    public ModelAndView willAdd() {
        ModelMap map = new ModelMap();
        List<RcCompany> companylist = rcCompanyService.list();
        return new ModelAndView("rc_user/add", "companylist", companylist);
    }
    @ApiOperation(value = "添加")
    @PostMapping("/add")
    @Transactional
    public String add(RcUser rcUser) {
        rcUser.setUserPwd(passwordEncoder.encode(rcUser.getUserPwd()));
        rcUserService.addUser(rcUser);
        return "redirect:/rcUser/manage";
    }
    @ApiOperation(value = "个人要修改")
    @GetMapping("/personalWillEdit")
    public ModelAndView personalWillEdit() {
```

```java
        RcUser rcUser = this.getCurrentUser();
        return new ModelAndView("rc_user/personalEdit", "rcUser", rcUser);
    }
    @ApiOperation(value = "个人修改")
    @PostMapping("/personalEdit")
    @Transactional
    public ModelAndView personalEdit(RcUser rcUser) {
        RcUser user1 = rcUserService.getById(rcUser.getUserId());
        rcUser.setUserPwd(user1.getUserPwd());
        return new ModelAndView("common/message", "msg", "修改成功！");
    }
    @ApiOperation(value = "要修改")
    @GetMapping("/willEdit")
    public ModelAndView willEdt(Integer userId) {
        ModelMap map = new ModelMap();
        List<RcCompany> companylist = rcCompanyService.list();
        RcUser rcUser = rcUserService.getById(userId);
        map.put("companylist", companylist);
        map.put("rcUser", rcUser);
        return new ModelAndView("rc_user/edit", "map", map);
    }
    @ApiOperation(value = "修改")
    @PostMapping("/edit")
    @Transactional
    public String edit(RcUser rcUser) {
        RcUser user1 = rcUserService.getById(rcUser.getUserId());
        rcUser.setUserPwd(user1.getUserPwd());
        rcUserService.editUser(rcUser);
        return "redirect:/rcUser/manage";
    }
    @ApiOperation(value = "要找回密码")
    @GetMapping("/willFindPwd")
    public String willFindPwd() {
        return "rc_user/findPwd";
    }
    @ApiOperation(value = "要找回密码")
    @PostMapping("/forFindPwd")
    public ModelAndView forFindPwd(String userName, String userEmail, HttpServletRequest request) {
        QueryWrapper<RcUser> queryWrapper = new QueryWrapper<>();
        queryWrapper.eq("user_name", userName);
        queryWrapper.eq("user_email", userEmail);
        RcUser rcUser = rcUserService.getOne(queryWrapper);
        if (rcUser != null) {
            return new ModelAndView("rc_user/setNewPwd", "userId", rcUser.getUserId());
        } else {
            String msg = "输入的信息有误！<a href='JavaScript:window.history.back()'>返回</a>";

            return new ModelAndView("common/message", "msg", msg);
        }
```

```java
}
@ApiOperation(value = "要设置密码")          //在找回密码时使用
@GetMapping("/willSetNewPwd/{id}")
public ModelAndView willSetNewPwd(@PathVariable("id") Integer id) {
    return new ModelAndView("rc_user/setNewPwd", "userId", id);
}
@ApiOperation(value = "设置密码")            //在找回密码时使用
@PostMapping("/setNewPwd")
public String setNewPwd(Integer userId, String newPwd) {
    RcUser rcUser = rcUserService.getById(userId);
    rcUser.setUserPwd(passwordEncoder.encode(newPwd));
    rcUserService.editUser(rcUser);
    return "redirect:/";
}
@ApiOperation(value = "要修改密码")          //在找回密码时使用
@GetMapping("/willEditPwd")
public ModelAndView willEditPwd() {
    RcUser rcUser = this.getCurrentUser();
    return new ModelAndView("rc_user/editPwd", "userId", rcUser.getUserId());
}
@ApiOperation(value = "修改密码")            //在找回密码时使用
@PostMapping("/editPwd")
public ModelAndView editPwd(Integer userId, String userPwd, String newPwd) {
    RcUser rcUser = rcUserService.getById(userId);
    boolean f = passwordEncoder.matches(userPwd, rcUser.getUserPwd());
    if (f) {
        rcUser.setUserPwd(passwordEncoder.encode(newPwd));
        rcUserService.editUser(rcUser);
        return new ModelAndView("common/message", "msg", "修改成功！");
    } else {
        return new ModelAndView("common/message", "msg", "原密码错误！修改失败！
        <a href='JavaScript:history.back()'>返回</a>");
    }
}
@ApiOperation(value = "根据ID删除")
@GetMapping("/delete/{id}")
@Transactional
public String delete(@PathVariable("id") Integer id) {
    rcUserService.removeById(id);
    return "redirect:/rcUser/manage";
}
@ApiOperation(value = "批量删除")
@PostMapping("/delete/batch")
@Transactional
public String deleteBatch(Integer[] ids) {
    List<Integer> idList = new ArrayList<Integer>();
    for (Integer id : ids) {
        idList.add(id);
    }
```

```java
        rcUserService.removeByIds(idList);
        return "redirect:/rcUser/manage";
    }
    @ApiOperation(value = "分页浏览")
    @RequestMapping("/browse")
    public ModelAndView browse(Integer pageNo, Integer pageSize) {
        if (pageNo == null) pageNo = 1;
        if (pageSize == null) pageSize = 10;
        ModelMap map = new ModelMap();
        IPage<RcUser> page = rcUserService.page(new Page<>(pageNo, pageSize));
        PageList<RcUser> plist = new PageList<RcUser>(page.getRecords(), page.getTotal(), pageNo, pageSize);
        map.put("pageNo", pageNo);
        map.put("pageSize", pageSize);
        map.put("plist", plist);
        return new ModelAndView("rc_user/browse", "map", map);
    }
    @ApiOperation(value = "管理")
    @RequestMapping("/manage")
    public ModelAndView manage(String userName, Integer pageNo, Integer pageSize) {
        if (pageNo == null) pageNo = 1;
        if (pageSize == null) pageSize = 10;
        ModelMap map = new ModelMap();
        QueryWrapper<RcUser> queryWrapper = new QueryWrapper<>();
        if (StringUtils.isNotBlank(userName)) {
            queryWrapper.like("user_name", "%" + userName + "%");
        }
        IPage<RcUser> page = rcUserService.page(new Page<RcUser>(pageNo, pageSize), queryWrapper);
        PageList<RcUser> plist = new PageList<RcUser>(page.getRecords(), page.getTotal(), pageNo, pageSize);
        map.put("userName", userName);
        map.put("pageNo", pageNo);
        map.put("pageSize", pageSize);
        map.put("plist", plist);
        return new ModelAndView("rc_user/manage", "map", map);
    }
    @ApiOperation(value = "查看")
    @GetMapping("/show/{id}")
    public ModelAndView show(@PathVariable("id") Integer id) {
        RcUser rcUser = rcUserService.getById(id);
        return new ModelAndView("rc_user/show", "rcUser", rcUser);
    }
    @ApiOperation(value = "要修改角色")
    @GetMapping("/willEditRole/{id}")
    public ModelAndView willEditRole(@PathVariable("id") Integer id) {
        RcUser rcUser = rcUserService.getById(id);
        List<RcRole> list = rcRoleService.list();
        ModelMap map = new ModelMap();
```

```java
            map.put("userId", id);
            for (RcRole r : list) {
                r.setIsSelected(false);
                for (RcRole r1 : rcUser.getRoles()) {
                    if (r.getRoleId() == r1.getRoleId()) {
                        r.setIsSelected(true);
                        break;
                    }
                }
            }
            map.put("list", list);
            return new ModelAndView("rc_user/editRole", "map", map);
    }
    @ApiOperation(value = "修改角色")
    @PostMapping("/editRole/{id}")
    @Transactional
    public String editRole(@PathVariable("id") Integer id, Integer[] roleIds) {
        rcUserService.editRole(id, roleIds);
        return "redirect:/rcUser/manage";
    }
}
```

2. 角色控制类

角色控制类的代码如下：

```java
@Slf4j
@Api(tags = {"角色控制类"})
@Controller
@RequestMapping("/rcRole")
public class RcRoleController extends RcBaseController {
    @Resource
    public RcRoleService rcRoleService;
    @Resource
    public RcPermissionService rcPermissionService;
    @ApiOperation(value = "要添加")
    @GetMapping("/willAdd")
    public String willAdd(){
        return "rc_role/add";
    }
    @ApiOperation(value = "添加")
    @PostMapping("/add")
    @Transactional
    public String add(RcRole rcRole){
        rcRoleService.save(rcRole);
        return "redirect:/rcRole/manage";
    }
    @ApiOperation(value = "要修改")
    @GetMapping("/willEdit/{id}")
    public ModelAndView willEdt(@PathVariable("id") Integer id){
        RcRole rcRole = rcRoleService.getById(id);
```

```java
        return new ModelAndView("rc_role/edit","rcRole",rcRole);
    }
    @ApiOperation(value = "修改")
    @PostMapping("/edit")
    public String edit(RcRole rcRole){
        rcRoleService.updateById(rcRole);
        return "redirect:/rcRole/manage";
    }
    @ApiOperation(value = "根据ID删除")
    @GetMapping("/delete/{id}")
    public String delete(@PathVariable("id") Integer id){
        rcRoleService.removeById(id);
        return "redirect:/rcRole/manage";
    }
    @ApiOperation(value = "管理")
    @RequestMapping("/manage")
    public ModelAndView manage(Integer pageNo, Integer pageSize){
        if(pageNo==null)pageNo=1;
        if(pageSize==null)pageSize=10;
        ModelMap map = new ModelMap();
        IPage<RcRole> page = rcRoleService.page(new Page<>(pageNo, pageSize));
        PageList<RcRole> plist = new PageList<RcRole>(page.getRecords(), page.getTotal(), pageNo, pageSize);
        map.put("pageNo", pageNo);
        map.put("pageSize", pageSize);
        map.put("plist", plist);
        return new ModelAndView("rc_role/manage","map",map);
    }
    @ApiOperation(value = "要修改权限")
    @GetMapping("/willEditPermission/{id}")
    public ModelAndView willEditPermission(@PathVariable("id") Integer id){
        RcRole rcRole = rcRoleService.getById(id);
        List<RcPermission> list = rcPermissionService.list();
        ModelMap map=new ModelMap();
        map.put("roleId",id);
        for(RcPermission p:list){
            p.setIsSelected(false);
            for(RcPermission p1:rcRole.getPermissions()){
                if(p.getPermId()==p1.getPermId()){
                    p.setIsSelected(true);
                    break;
                }
            }
        }
        map.put("list",list);
        return new ModelAndView("rc_role/editPermission","map",map);
    }
    @ApiOperation(value = "修改权限")
    @PostMapping("/editPermission/{id}")
```

```java
    public String editPermission(@PathVariable("id") Integer id,Integer[] permIds){
        rcRoleService.editPermissions(id,permIds);
        return "redirect:/rcRole/manage";
    }
}
```

3. 权限控制类

权限控制类的代码如下:

```java
@Slf4j
@Api(tags = {"权限控制类"})
@Controller
@RequestMapping("/rcPermission")
public class RcPermissionController extends RcBaseController {
    @Resource
    public RcPermissionService rcPermissionService;
    @ApiOperation(value = "要添加")
    @GetMapping("/willAdd")
    public String willAdd(){
        return "rc_permission/add";
    }
    @ApiOperation(value = "添加")
    @PostMapping("/add")
    public String add(RcPermission rcPermission){
        rcPermissionService.save(rcPermission);
        return "redirect:/rcPermission/manage";
    }
    @ApiOperation(value = "要修改")
    @GetMapping("/willEdit/{id}")
    public ModelAndView willEdt(@PathVariable("id") Integer id){
        RcPermission rcPermission = rcPermissionService.getById(id);
        return new ModelAndView("rc_permission/edit","rcPermission",rcPermission);
    }
    @ApiOperation(value = "修改")
    @PostMapping("/edit")
    public String edit(RcPermission rcPermission){
        rcPermissionService.updateById(rcPermission);
        return "redirect:/rcPermission/manage";
    }
    @ApiOperation(value = "根据id删除")
    @GetMapping("/delete/{id}")
    public String delete(@PathVariable("id") Integer id){
        rcPermissionService.removeById(id);
        return "redirect:/rcPermission/manage";
    }
    @ApiOperation(value = "管理")
    @RequestMapping("/manage")
    public ModelAndView manage(Integer pageNo, Integer pageSize){
        if(pageNo==null)pageNo=1;
        if(pageSize==null)pageSize=10;
```

```java
        ModelMap map = new ModelMap();
        IPage<RcPermission> page = rcPermissionService.page(new Page<>(pageNo, pageSize));
        PageList<RcPermission> plist = new PageList<RcPermission>(page.getRecords(),
page.getTotal(), pageNo, pageSize);
        map.put("pageNo", pageNo);
        map.put("pageSize", pageSize);
        map.put("plist", plist);
       return new ModelAndView("rc_permission/manage","map",map);
    }
}
```

5.2.5　权限相关视图层设计

1．权限视图设计

1）权限管理

【权限管理】界面如图 5-13 所示。管理员可以在此分页浏览权限，或者添加、修改、删除权限。

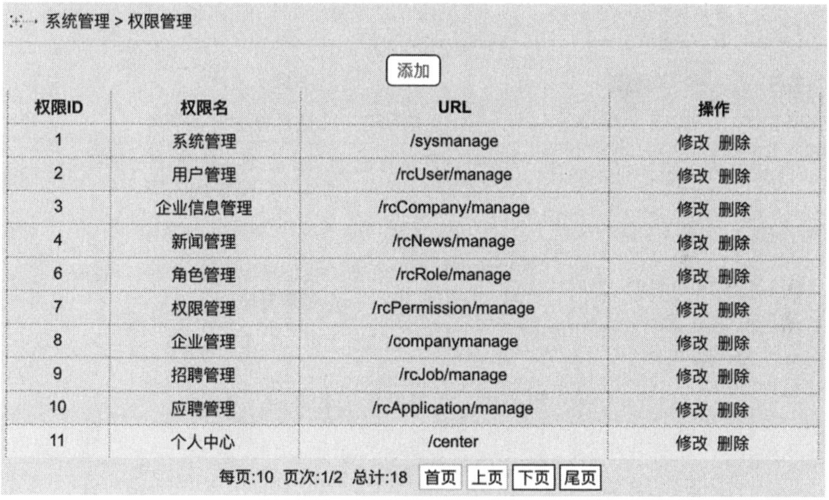

权限ID	权限名	URL	操作
1	系统管理	/sysmanage	修改 删除
2	用户管理	/rcUser/manage	修改 删除
3	企业信息管理	/rcCompany/manage	修改 删除
4	新闻管理	/rcNews/manage	修改 删除
6	角色管理	/rcRole/manage	修改 删除
7	权限管理	/rcPermission/manage	修改 删除
8	企业管理	/companymanage	修改 删除
9	招聘管理	/rcJob/manage	修改 删除
10	应聘管理	/rcApplication/manage	修改 删除
11	个人中心	/center	修改 删除

图 5-13　【权限管理】界面

2）添加权限

【添加权限】界面如图 5-14 所示。

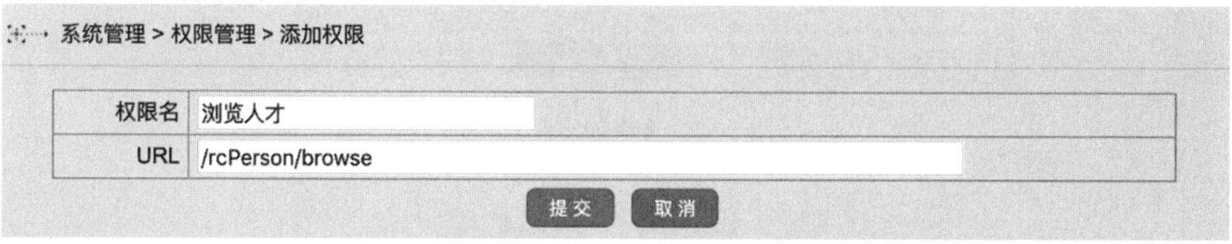

图 5-14　【添加权限】界面

3）修改权限

【修改权限】界面如图 5-15 所示。

图 5-15 【修改权限】界面

2. 角色视图设计

1）角色管理

【角色管理】界面如图 5-16 所示。管理员可以在此分页浏览角色，或者添加、修改、删除角色，或者设置角色具有的权限。

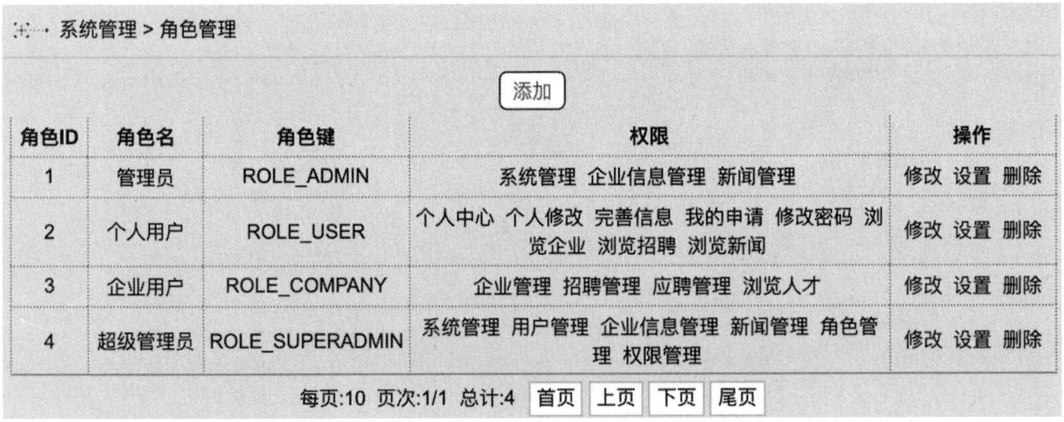

图 5-16 【角色管理】界面

2）添加角色

【添加角色】界面如图 5-17 所示。

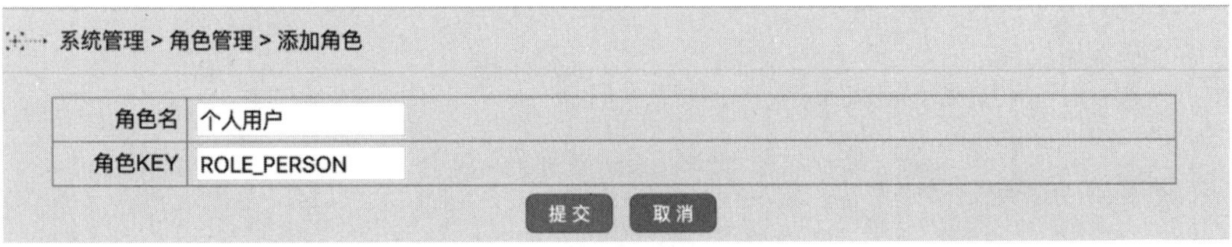

图 5-17 【添加角色】界面

3）修改角色

【修改角色】界面如图 5-18 所示。

图 5-18　【修改角色】界面

4）设置权限

【设置权限】界面如图 5-19 所示。

选择	权限ID	权限名	URL
☑	1	系统管理	/sysmanage
☑	2	用户管理	/rcUser/manage
☑	3	企业信息管理	/rcCompany/manage
☑	4	新闻管理	/rcNews/manage
☑	6	角色管理	/rcRole/manage
☑	7	权限管理	/rcPermission/manage
☐	8	企业管理	/companymanage
☐	9	招聘管理	/rcJob/manage
☐	10	应聘管理	/rcApplication/manage
☐	11	个人中心	/center
☐	12	个人修改	/rcUser/personalWillEdit
☐	13	完善信息	/rcPerson/willSetInfo
☐	14	我的申请	/rcApplication/browse
☐	15	修改密码	/rcUser/willEditPwd
☐	16	浏览企业	/rcCompany/browse
☐	17	浏览招聘	/rcJob/browse
☐	19	浏览人才	/rcPerson/browse

图 5-19　【设置权限】界面

3．用户视图设计

1）用户管理

【用户管理】界面如图 5-20 所示。管理员可以在此分页浏览用户，或者添加、修改、删除和查询用户，或者设置用户角色。

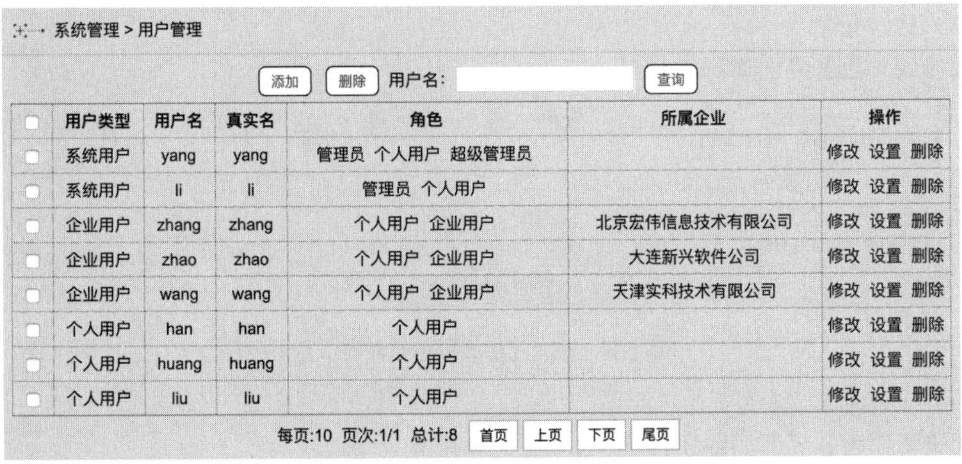

图 5-20 【用户管理】界面

2）添加用户

【添加用户】界面如图 5-21 和图 5-22 所示。

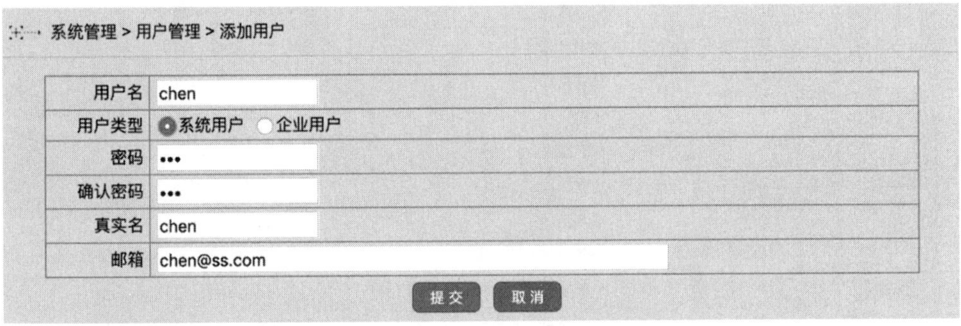

图 5-21 【添加用户】界面（系统用户）

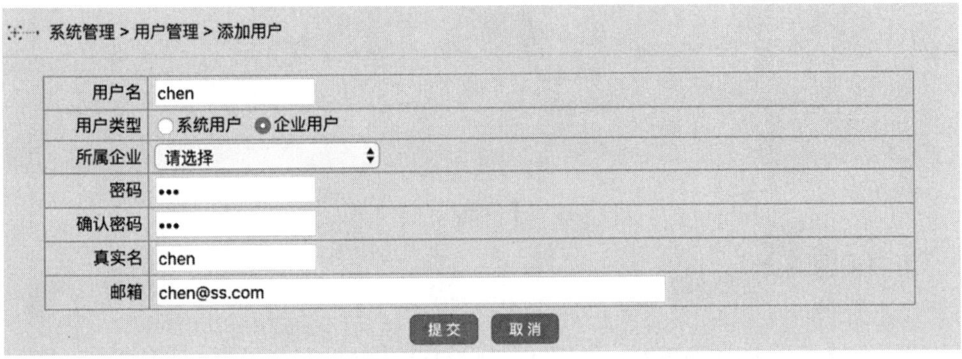

图 5-22 【添加用户】界面（企业用户）

3）修改用户

【修改用户】界面如图 5-23 所示。

图 5-23　【修改用户】界面

4）设置角色

【设置角色】界面如图 5-24 所示。

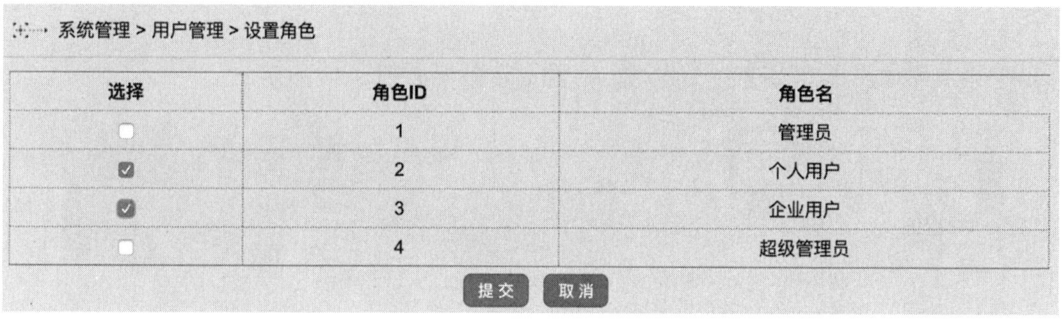

图 5-24　【设置角色】界面

5）注册用户

【注册用户】界面如图 5-25 所示。

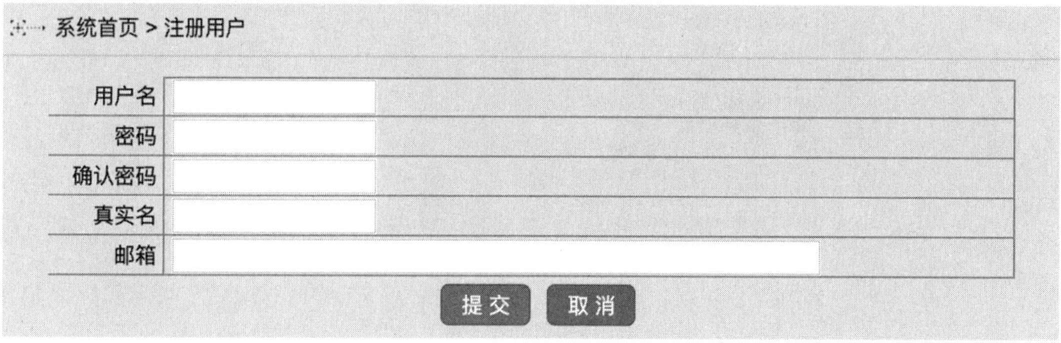

图 5-25　【注册用户】界面

6）找回密码

【找回密码】界面如图 5-26 和图 5-27 所示。

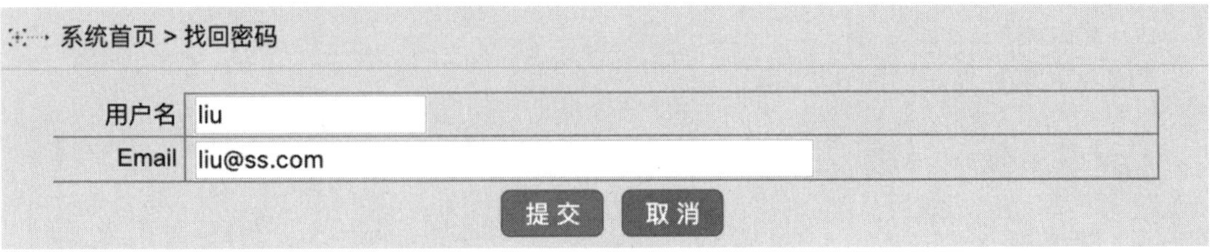

Java EE 企业级应用开发技术研究

图 5-26 【找回密码】界面

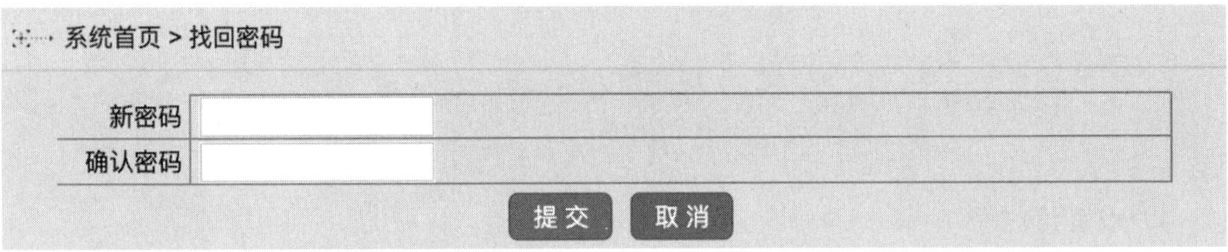

图 5-27 【找回密码】界面（设置新密码）

7）登录

登录界面如图 5-28 所示，登录成功后的显示效果如图 5-29 所示。

图 5-28 登录界面

第 5 章 基于 Spring Security 实现认证和授权

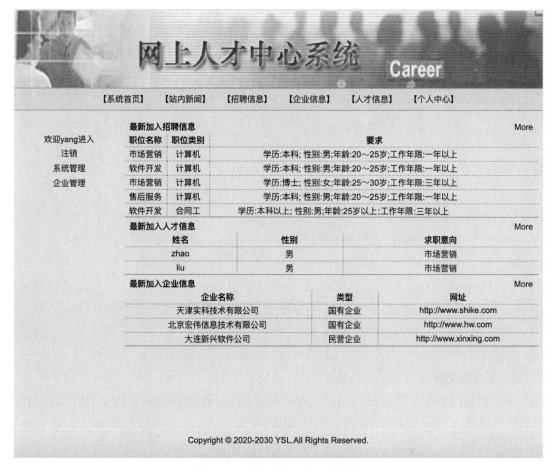

图 5-29 登录成功后的显示效果

5.3 权限相关组件设计及其配置设计

5.3.1 权限相关组件设计

1. 决策管理器类

决策管理器类是授权的主要类，用于决定用户是否具有访问当前 URL 的权限。例如：

```
@Component
public class RcAccessDecisionManager  implements AccessDecisionManager {
    // decide()方法是判定是否拥有权限的决策方法
    @Override
    public void decide(Authentication authentication, Object object, Collection<ConfigAttribute> configAttributes) throws AccessDeniedException, InsufficientAuthenticationException {
        if (null == configAttributes || configAttributes.size() <= 0) {
            return;
        }
        ConfigAttribute c;
```

```
            String needRole;
            for (Iterator<ConfigAttribute> iter = configAttributes.iterator(); iter.hasNext(); ) {
                c = iter.next();
                needRole = c.getAttribute();
                for (GrantedAuthority ga : authentication.getAuthorities()) {
                    if (needRole.trim().equals(ga.getAuthority())) {
                        return;
                    }
                }
            }
            throw new AccessDeniedException("权限不足！");
        }。
        @Override
        public boolean supports(ConfigAttribute attribute) {
            return true;
        }
        @Override
        public boolean supports(Class<?> clazz) {
            return true;
        }
}
```

2. 安全元数据类

安全元数据类实现了 FilterInvocationSecurityMetadataSource 接口，用于获取当前 URL 所对应的权限，并提供给权限决策管理器使用。例如：

```
@Component
public class RcInvocationSecurityMetadataSourceService implements
        FilterInvocationSecurityMetadataSource {
    @Resource
    private RcPermissionService permissionService;
    private HashMap<String, Collection<ConfigAttribute>> map =null;
    /**
     * 加载权限表中的所有权限
     */
    public void loadResourceDefine(){
        map = new HashMap<>();
        Collection<ConfigAttribute> array;
        List<RcPermission> permissions = permissionService.list();
        for(RcPermission permission : permissions) {
            array = new ArrayList<>();
            for(RcRole role : permission.getRoles()) {
                array.add(new SecurityConfig(role.getRoleKey()));
            }
            map.put(permission.getPermUrl(), array);
        }
    }
```

```java
//此方法是为了判定用户请求的URL是否在权限表中。如果URL在权限表中,则将该URL的权限
//返回给decide()方法,用于判定用户是否具有此权限;如果URL不在权限表中,则放行
@Override
public Collection<ConfigAttribute> getAttributes(Object object) throws IllegalArgumentException {
    if(map ==null) loadResourceDefine();
    HttpServletRequest request = ((FilterInvocation) object).getHttpRequest();
    AntPathRequestMatcher matcher;
    String resUrl;
    for(Iterator<String> iter = map.keySet().iterator(); iter.hasNext(); ) {
        resUrl = iter.next();
        matcher = new AntPathRequestMatcher(resUrl);
        if(matcher.matches(request)) {
            return map.get(resUrl);
        }
    }
    return null;
}
@Override
public Collection<ConfigAttribute> getAllConfigAttributes() {
    return null;
}
@Override
public boolean supports(Class<?> clazz) {
    return FilterInvocation.class.isAssignableFrom(clazz);
}
}
```

3. 权限过滤器类

权限过滤器类继承于 AbstractSecurityInterceptor 类,并实现了 Filter 接口,用于拦截请求,将自定义的权限数据源类与自定义的权限决策管理器类整合在一起实现授权决策。例如:

```java
@Component
public class RcFilterSecurityInterceptor extends AbstractSecurityInterceptor implements Filter {
    @Resource
    private FilterInvocationSecurityMetadataSource mySecurityMetadataSource;
    @Resource
    @Override
    public void setAccessDecisionManager(AccessDecisionManager myAccessDecisionManager) {
        super.setAccessDecisionManager(myAccessDecisionManager);
    }
    @Override
    public SecurityMetadataSource obtainSecurityMetadataSource() {
        return this.mySecurityMetadataSource;
    }
    @Override
    public void init(FilterConfig filterConfig) throws ServletException {
    }
    @Override
```

```java
        public void doFilter(ServletRequest request, ServletResponse response, FilterChain chain) throws IOException, ServletException {
            FilterInvocation fi = new FilterInvocation(request, response, chain);
            invoke(fi);
        }
        public void invoke(FilterInvocation fi) throws IOException, ServletException {
            //fi 中有一个被拦截的 URL
            //调用 MyInvocationSecurityMetadataSource 的 getAttributes(Object object)方法获取 fi 对应的
            //所有权限
            //再调用 MyAccessDecisionManager 的 decide()方法来校验用户的权限是否足够
            InterceptorStatusToken token = super.beforeInvocation(fi);
            try {
                //执行下一个拦截器
                fi.getChain().doFilter(fi.getRequest(), fi.getResponse());
            } finally {
                super.afterInvocation(token, null);
            }
        }
        @Override
        public void destroy() {
        }
        @Override
        public Class<?> getSecureObjectClass() {
            return FilterInvocation.class;
        }
    }
```

4. 登录失败处理器类

登录失败处理器类继承于 SimpleUrlAuthenticationFailureHandler 类，代码如下：

```java
@Slf4j
public class RcAuthenctiationFailureHandler extends SimpleUrlAuthenticationFailureHandler {
    public RcAuthenctiationFailureHandler(String defaultFailureUrl) {
        super(defaultFailureUrl);
    }
    @Autowired
    private ObjectMapper objectMapper;
    @Override
    public void onAuthenticationFailure(HttpServletRequest request, HttpServletResponse response,
            AuthenticationException exception) throws IOException, ServletException {
        log.info("登录失败");
        super.onAuthenticationFailure(request,response,exception);
    }
}
```

5. 授权不通过处理器类

授权不通过处理器类继承于 SimpleUrlAuthenticationFailureHandler 类，代码如下：

```java
@Slf4j
public class RcAuthenctiationFailureHandler extends SimpleUrlAuthenticationFailureHandler {
```

```java
    private String defaultFailureUrl;
    public RcAuthenctiationFailureHandler(String defaultFailureUrl) {
        super(defaultFailureUrl);
        this.defaultFailureUrl = defaultFailureUrl;
    }
    @Autowired
    private ObjectMapper objectMapper;
    @Override
    public void onAuthenticationFailure(HttpServletRequest request, HttpServletResponse response,
AuthenticationException exception) throws IOException, ServletException {
        log.info("登录失败");
        super.onAuthenticationFailure(request,response,exception);
    }
}
```

5.3.2 验证码实现相关设计

1．验证码的一些常量定义

```java
public class RcConstants {
    public static final String SESSION_KEY = "SESSION_KEY_IMAGE_CODE";
    //图片宽度
    public static final int WIDTH = 50;
    //图片高度
    public static final int HEIGHT = 20;
    //验证码的位数
    public static final int RANDOM_SIZE = 4;
    //验证码过期秒数
    public static final int EXPIRE_SECOND = 30;
}
```

2．验证码异常类设计

```java
public class RcValidateCodeException extends AuthenticationException {
    public RcValidateCodeException(String msg, Throwable t) {
        super(msg, t);
    }
    public RcValidateCodeException(String msg) {
        super(msg);
    }
}
```

3．验证码数据模型类设计

```java
@Data
@AllArgsConstructor
public class ImageCode {
    private BufferedImage image;
    private String code;
    private LocalDateTime expireTime;
    public boolean isExpried() {
```

```
        return LocalDateTime.now().isAfter(expireTime);
    }
}
```

4. 生成验证码的控制器类设计

```
@Controller
@RequestMapping("/code")
public class RcValidateCodeController {
    @Resource
    private SessionStrategy sessionStrategy;
    @GetMapping("/image")
    public void createCode(HttpServletRequest request, HttpServletResponse response) throws IOException {
        response.setContentType("image/jpeg");
        ImageCode imageCode = generate(new ServletWebRequest(request));
        sessionStrategy.setAttribute(new ServletWebRequest(request), RcConstants.SESSION_KEY, imageCode);
        ImageIO.write(imageCode.getImage(), "JPEG", response.getOutputStream());
    }
    private ImageCode generate(ServletWebRequest request) {
        int width = ServletRequestUtils.getIntParameter(request.getRequest(), "width", RcConstants.WIDTH);
        int height = ServletRequestUtils.getIntParameter(request.getRequest(), "height", RcConstants.HEIGHT);
        BufferedImage image = new BufferedImage(width, height, BufferedImage.TYPE_INT_RGB);
        Graphics2D g = image.createGraphics();
        g.setRenderingHint(RenderingHints.KEY_TEXT_ANTIALIASING,
                RenderingHints.VALUE_TEXT_ANTIALIAS_LCD_HRGB);  //消除锯齿状
        Random random = new Random();
        g.setColor(Color.GRAY);
        g.fillRect(0, 0, width, height);
        g.setColor(Color.white);
        g.setFont(new Font("Arial", Font.PLAIN, 14));
        String sRand = "";
        int length = ServletRequestUtils.getIntParameter(request.getRequest(), "length", RcConstants.RANDOM_SIZE);
        for (int i = 0; i < length; i++) {
            String rand = String.valueOf(random.nextInt(10));
            sRand += rand;
            g.drawString(rand, 10 * i+4 , 16);
        }
        g.dispose();
        LocalDateTime ldt=LocalDateTime.now().plusSeconds(60);
        return new ImageCode(image, sRand, ldt);
    }
}
```

5. 用于判断验证码的过滤器设计

```
@Slf4j
```

```java
@Component("validateCodeFilter")
public class RcValidateCodeFilter extends OncePerRequestFilter implements Filter {
    @Resource
    private AuthenticationFailureHandler authenticationFailureHandler;
    @Resource
    private SessionStrategy sessionStrategy;
    @Override
    protected void doFilterInternal(HttpServletRequest request, HttpServletResponse response,
FilterChain filterChain)throws ServletException, IOException {
        log.info("request : {}", request.getRequestURI());
        //必须是登录的Post请求才会进行验证,其他的直接放行
        if(StringUtils.equals("/rc/sign-in", request.getRequestURI()) &&
            StringUtils.equalsIgnoreCase(request.getMethod(), "post")) {
            log.info("request : {}", request.getRequestURI());
            try {
                //进行验证码的校验
                validate(new ServletWebRequest(request));
            } catch (AuthenticationException e) {
                authenticationFailureHandler.onAuthenticationFailure(request,response,e);
                return;
            }
        }
        //如果校验通过,就放行
        filterChain.doFilter(request, response);
    }
    private void validate(ServletWebRequest request) throws ServletRequestBindingException {
        //获取请求中的验证码
        String codeInRequest = ServletRequestUtils.getStringParameter(request.getRequest(),
"imageCode");
        //校验空值情况
        if(StringUtils.isEmpty(codeInRequest)) {
            throw new ValidateCodeException("验证码不能为空");
        }
        //获取服务器session池中的验证码
        ImageCode codeInSession = (ImageCode) sessionStrategy.getAttribute(request,
RcConstants.SESSION_KEY);
        if(Objects.isNull(codeInSession)) {
            throw new ValidateCodeException("验证码不存在");
        }
        //校验服务器session池中的验证码是否过期
        if(codeInSession.isExpried()) {
            sessionStrategy.removeAttribute(request, RcConstants.SESSION_KEY);
            throw new ValidateCodeException("验证码过期了");
        }
        //请求验证码校验
        if(!StringUtils.equals(codeInSession.getCode(), codeInRequest)) {
            throw new ValidateCodeException("验证码不匹配");
        }
```

```
        //移除已完成校验的验证码
        sessionStrategy.removeAttribute(request, RcConstants.SESSION_KEY);
    }
}
```

5.3.3 权限相关配置设计

1. 安全配置

```
@Configuration
@EnableWebSecurity
public class RcSecurityConfiguration extends WebSecurityConfigurerAdapter {
    //实际没必要
    @Resource
    private RcUserService userService;
    @Resource
    private RcValidateCodeFilter validateCodeFilter;
    @Resource
    private RcFilterSecurityInterceptor filterSecurityInterceptor;
    @Bean
    public AuthenticationFailureHandler authenctiationFailureHandler(){
        return new RcAuthenctiationFailureHandler("/");
    }
    @Bean
    public PasswordEncoder passwordEncoder(){
        return new BCryptPasswordEncoder();
    }
    @Override
    public void configure(WebSecurity web) throws Exception {
        web.ignoring().antMatchers("/static/**/*.*");
        web.ignoring().antMatchers("/v2/api-docs",
                "/swagger-resources/configuration/ui",
                "/swagger-resources",
                "/swagger-resources/configuration/security",
                "/swagger-ui.html"
        );
    }
    //定义认证规则
    @Override
    protected void configure(AuthenticationManagerBuilder auth) throws Exception {
        auth.userDetailsService(userService).passwordEncoder(passwordEncoder());
    }
    //定义授权规则
    @Override
    protected void configure(HttpSecurity http) throws Exception {
        http.cors().disable()               //关闭跨域访问控制
            .csrf().disable()               //关闭 CSRF
            .headers().frameOptions().sameOrigin();  //允许同域框架
        http.formLogin()
```

```
                .loginPage("/")                          //设置登录页的路径
                .usernameParameter("userName")           //指定用户名域名称,默认为username
                .passwordParameter("userPwd")            //指定密码域名称,默认为password
                .loginProcessingUrl("/sign-in")          //自定义登录验证接口,默认为/login
                .failureHandler(authenctiationFailureHandler())
                .defaultSuccessUrl("/manage")            //登录成功后跳转到首页
                .permitAll();
             http.logout()
                  .clearAuthentication(true)
                  .logoutSuccessUrl("/")
                  .invalidateHttpSession(true)
                  .permitAll();
            //定制请求授权规则
            http.authorizeRequests()
                  .antMatchers("/","/**/*").permitAll()
                  .and()
                  .exceptionHandling()
                  .accessDeniedHandler(new RcAccessDeniedHandler("/WEB-INF/jsp/common/
                   message.jsp"));
            //配置过滤器
            http.addFilterBefore(validateCodeFilter, UsernamePasswordAuthenticationFilter.class);
            http.addFilterBefore(filterSecurityInterceptor, FilterSecurityInterceptor.class);
        }
        @Bean
        public SessionStrategy registBean() {
            SessionStrategy sessionStrategy = new HttpSessionSessionStrategy();
            return sessionStrategy;
        }
}
```

2．国际化配置

为了用中文显示提示信息,可采用国际化技术。

(1) 在 resources 文件夹下建立 messages_zh_CN.properties 文件,内容如下:

```
AbstractAccessDecisionManager.accessDenied=不允许访问
AbstractSecurityInterceptor.authenticationNotFound=未在 SecurityContext 中找到认证对象
AbstractUserDetailsAuthenticationProvider.badCredentials=用户名或密码错误
AbstractUserDetailsAuthenticationProvider.credentialsExpired=用户凭证已过期
AbstractUserDetailsAuthenticationProvider.disabled=用户已失效
AbstractUserDetailsAuthenticationProvider.expired=用户账号已过期
AbstractUserDetailsAuthenticationProvider.locked=用户账号已被锁定
```

(2) 修改主类,增加 MessageSource 配置,代码如下:

```
@SpringBootApplication
@EnableTransactionManagement
public class RcApplication {
    public static void main(String[] args) {
        SpringApplication.run(RcApplication.class, args);
    }
    @Bean
```

```
    public MessageSource messageSource() {
        Locale.setDefault(Locale.CHINA);
        ReloadableResourceBundleMessageSource messageSource = new ReloadableResourceBundleMessageSource();
        messageSource.setDefaultEncoding("UTF-8");
        messageSource.addBasenames("classpath:messages_zh_CN");
        return messageSource;
    }
}
```

第 6 章 微服务架构与 Spring Cloud

本章主要介绍微服务架构的概念、Spring Cloud 核心技术，并通过网上人才中心系统的实现完成微服务架构设计。

6.1 微服务架构概述

随着互联网的快速发展，云计算在近十年得到了蓬勃发展，企业的 IT 环境和 IT 架构也逐渐发生变革，从过去的单体架构发展为至今广泛流行的微服务架构。微服务是一种架构风格，能够给软件应用开发带来很大的便利，但微服务的实施和落地面临很大的挑战，因此需要一套完整的微服务解决方案。在 Java 领域，Spring 框架的出现给 Java 企业级应用开发带来了福音，提高了开发效率。2014 年年底，Spring 团队推出 Spring Cloud，目标是使其成为 Java 领域微服务架构落地的标准。发展至今，Spring Cloud 已经成为 Java 领域微服务架构落地的完整解决方案，为企业 IT 架构的变革保驾护航。

6.1.1 单体架构与微服务架构

1. 单体架构

将所有的代码及功能都包含在一个归档包（可以是 jar、war、ear 或其他归档格式）中的项目组织方式被称为单体架构，如图 6-1 所示。

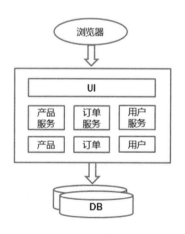

图 6-1 单体架构

对于 Java Web 应用而言，当采用单体架构时，一个项目通常会由多个模块组成，并且各模块会根据自身所提供的功能具有一个明确的边界。在编译时，这些模块会被打包成为一个个 jar 包，并最终合并在一起形成一个 war 包。然后该 war 包会被上传到 Web 服务器。在启动 Web 服务器后，Web 容器将自动解压 war 包。

单体架构的优点如下所述。
- 功能层次拆分清楚：单体架构的应用有着良好的功能层次划分，非常适合项目初期不太明确项目功能时的开发。
- 部署简单：单体架构是完整的结构体，可以直接部署在一个服务器上。
- 技术单一：项目不需要复杂的技术栈，往往使用一套熟悉的技术栈就可以完成开发。
- 用人成本低：单个程序员可以完成从业务接口到数据库的整个流程。

单体架构的缺点如下所述。
- 维护麻烦：即使只需更改一行代码，也需要重新编译打包整个项目。如果项目的代码量多，则每次都会花费大量的时间来重新编译、测试和部署。
- 技术选型固定：在项目变得越来越庞大的同时，为了满足更多的需求，所使用的技术会越来越多，但有些技术之间是不兼容的。在这种情况下，我们需要放弃使用某些不兼容的技术，而选择使用一种不是非常适合的技术来实现特定的功能。
- 可扩展性差：按照单体架构的代码，最终形成一个包含了所有功能的 war 包，因此在对服务容量进行扩展时，只能重复地部署整体 war 包来扩展服务容量，而不能单独扩展出现瓶颈的某一个模块。也就是说，在单体应用中，多个模块的负载不均，会导致我们在扩容高负载的容器时，也将低负载的容器扩容了，极大地浪费了资源。
- 维护升级成本高：在项目的整个周期中，可能会伴随着开发人员、运维人员的变更，导致很难准确、迅速地达到目的。如果需要升级某个模块，则很可能需要修改整个项目，这会耗费大量的人力、财力。

目前的单体架构存在很多弊端，但由于前期用户量不多、产品迭代不是很频繁，因此相应的问题并没有凸显。随着团队越来越大，相应的沟通成本、管理成本、人员协调成本会显著增加。同时，单体架构引起缺陷的原因组合多，导致分析、定位、修复缺陷的成本相应增加；在自动化测试机制不够完善的情况下，也非常容易导致"修复越多，缺陷越多"的恶性循环。

2. 微服务架构

微服务架构是指开发一个个小型的但有业务功能的服务，并使每个服务都有自己的处理和轻量通信机制，可以部署在单个或多个服务器上的项目组织方式，如图 6-2 所示。微服务架构也指一种松耦合的、有一定的有界上下文的面向服务架构。也就是说，如果所有服务都需要同时修改，那么它们就不是微服务，因为它们是紧耦合的；如果用户需要掌握一个包括很多服务的上下文场景使用条件，那么它就是一个有上下文边界的面向服务架构。

微服务架构模式的目的是将大型的、复杂的、长期运行的应用程序构建为一组相互配合的服务，每个服务都可以很容易地进行局部改良。每个服务都应该足够"小"。但是，这里的"小"不应该从代码量上比较，而应该从业务逻辑上比较，符合 SRP（Single Responsibility

Principle，单一职责）原则的服务才叫微服务。

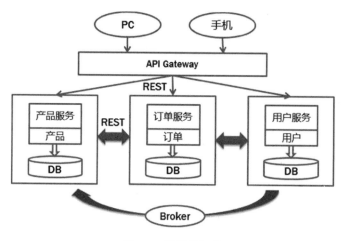

图 6-2 微服务架构

相对于单体架构和 SOA，微服务架构的主要特点是组件化、松耦合、自治和去中心化，这体现在以下几个方面。

- 一组小的服务：服务"粒度"小，每个服务是针对一个单一职责的业务能力的封装，专注于做好一件事情。
- 独立部署、运行和扩展：每个服务都能够被独立部署并运行在一个进程内。这种部署和运行方式能够赋予系统灵活的代码组织方式和发布节奏，使得快速交付和应对变化成为可能。
- 独立开发和演化：技术选型灵活，不受遗留系统技术约束。合适的业务问题选择合适的技术，可以实现独立演化。服务与服务之间采取与语言无关的 API 进行集成。相对于单体架构，微服务架构是一种面向业务创新的架构模式。
- 独立团队和自治：团队对服务的整个生命周期负责，工作在独立的上下文中，自己决策、自己治理，而不需要统一的指挥中心。团队和团队之间通过松散的社区部落进行衔接。

可以看到，整个微服务架构的思想就是通过解耦我们所做的事情，并分而治之，以减少不必要的损耗，使得整个复杂的系统和组织能够快速地应对变化。当然，微服务架构也有缺点，如下所述。

- 运维要求高：更多的服务意味着需要投入更多的运维成本。
- 分布式固有的复杂性：使用微服务架构构建的是分布式系统，而对于一个分布式系统而言，系统容错、网络延迟、分布式事务等都会面临巨大的问题。
- 接口调整成本高：微服务之间通过接口进行通信，如果修改某一个微服务的 API，则所有用到这个接口的微服务可能都需要调整。

6.1.2 Spring Cloud 概述

Spring Cloud 是一个基于 Spring Boot 实现的云原生应用开发工具，为基于 JVM 的云原生应用开发中涉及的配置管理、服务发现、熔断器、路由、微代理、事件总线、全局锁、精选决策、分布式会话等操作提供了一整套解决方案。

尽管 Spring Cloud 带有"Cloud"这个单词，但它并不是云计算解决方案，而是在 Spring Boot 基础上构建的、用于快速构建分布式系统的通用模式的工具集。然而使用 Spring Cloud 开发的应用程序非常适合在 Docker 和 PaaS（如 Pivotal Cloud Foundry）上部署，所以又叫作云原生应用（Cloud Native Application）。云原生可以被简单地理解为面向云环境的软件架构。

Spring Cloud 作为第二代微服务的代表性框架，已经在国内众多大中小型的公司中有实际应用案例。许多公司的业务线全部应用 Spring Cloud，部分公司的业务线选择部分应用 Spring Cloud。典型的微服务架构如图 6-3 所示。

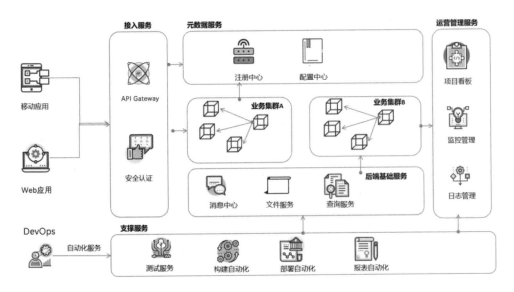

图 6-3 典型的微服务架构

Spring Cloud 与 Spring Boot 的不同在于：Spring Boot 专注于快速、方便地开发单体微服务，而 Spring Cloud 专注于全局的微服务，协调整理治理框架，它将 Spring Boot 开发的一个个单体微服务整合并管理起来，为各服务提供配置管理、服务发现、熔断器、路由、微代理、事件总线、全局锁、精选决策、分布式会话等集成服务；Spring Boot 可以离开 Spring Cloud 来独立开发项目，但 Spring Cloud 离不开 Spring Boot，二者属于单向依赖关系。

Spring Cloud Netflix 为 Spring Cloud 的核心套件，可以对多个 Netflix OSS 开源套件进行整合，包括以下组件。

- Eureka：服务治理组件，包含服务注册与发现。
- Hystrix：容错管理组件，实现了熔断器功能。

- Ribbon：客户端负载均衡的服务调用组件。
- Feign：基于 Ribbon 和 Hystrix 的声明式服务调用组件。
- Zuul：微服务网关组件，提供智能路由、访问过滤等功能。
- Archaius：外部化配置组件。
- Spring Cloud Config：配置管理工具，实现应用配置的外部化存储，支持客户端配置信息刷新、加密/解密配置内容等。
- Spring Cloud Bus：事件、消息总线，用于传播集群中的状态变化或事件，以及触发后续的处理。
- Spring Cloud Security：基于 Spring Security 的安全工具包，为应用程序添加安全控制功能。
- Spring Cloud Consul：封装了 Consul 操作。Consul 是一个服务发现与配置工具（与 Eureka 作用类似），可以与 Docker 容器无缝集成。

6.1.3　Spring Cloud 重要组件介绍

1．服务治理组件：Eureka

Eureka 是 Spring Cloud Netflix 微服务套件中的一部分，可以很容易地与 Spring Boot 构建的微服务整合起来。Eureka 包含了服务器端和客户端组件。服务器端，也称作服务注册中心，用于提供服务的注册与发现。Eureka 支持高可用的配置，当集群中有分片出现故障时，Eureka 会转入自动保护模式，它允许分片在故障期间继续提供服务的注册与发现；当故障分片恢复正常时，集群中其他分片会将它们的状态再次同步回来。客户端组件包含服务消费者与服务提供者。在应用程序运行时，Eureka 客户端向注册中心注册自身提供的服务并通过周期性地发送心跳来更新它的服务租约。同时，可以从服务器端查询当前注册的服务信息，将它们缓存到本地并周期性地刷新服务状态。

2．声明式服务调用组件：Feign

Feign 是一个声明式的伪 HTTP 客户端，它简化了服务消费者调用服务提供者的接口，使得编写 HTTP 客户端变得更加简单。在使用 Feign 时，只需要创建一个接口并通过注解的方式来配置它，即可完成对服务提供者的接口绑定，从而简化了在使用 Ribbon 时自行封装服务以调用客户端的代码开发量。

Feign 具有可插拔的注解特性，包括 Feign 注解和 JAX-RS 注解，同时扩展了对 Spring MVC 的注解支持。Feign 支持可插拔的编码器和解码器，默认集成了 Ribbon，并和 Eureka 结合，默认实现了负载均衡的效果。

3．客户端负载均衡的服务调用组件：Ribbon

Ribbon 是 Spring Cloud Netflix 发布的负载均衡器，它有助于控制 HTTP 和 TCP 客户端的行为。在为 Ribbon 配置服务提供者的地址后，Ribbon 就可以基于某种负载均衡算法自动地帮助服务消费者去请求服务。Ribbon 提供了很多负载均衡算法，如轮询、加权、随机等。当然，Ribbon 也可以实现自定义的负载均衡算法。

4. 容错管理组件：Hystrix

Hystrix 旨在通过控制服务和第三方库的节点对延迟及故障提供更加强大的容错能力。当被访问的微服务无法使用时，当前服务能够感知这个现象，并且提供一个备用的方案。

线程隔离：用户请求不直接访问服务，而是使用线程池中空闲的线程访问服务，以加速失败判断时间。

服务降级：及时返回服务调用失败的结果，让线程不因为等待服务而阻塞。

5. 微服务网关组件：Zuul

Zuul 是 Spring Cloud Netflix 的一个开源 API Gateway 服务器。Zuul 本质上是一个 Web Servlet 应用，用于在云平台上提供动态路由、监控、弹性、安全等边缘服务的框架。Zuul 相当于设备和 Netflix 流应用的 Web 网站后端所有请求的前门。

6.2 网上人才中心系统微服务工程设计

6.2.1 微服务设计基础

1. 微服务划分方法

微服务架构设计的首要任务是合理地划分微服务，即围绕业务功能来创建微服务。在进行微服务划分时，对于微服务粒度，建议在平台创建的初始阶段使用粗粒度的方法，按照业务功能进行划分；然后，随着业务的发展及平台运营情况的变化，按照发展规模考虑是否继续细分。下面将使用水平划分法和垂直划分法两种方法相结合的方式来创建微服务。

一方面，在水平方向上，按照业务功能的不同来划分微服务，并将这次划分所创建的微服务称为业务微服务。业务微服务负责业务功能的行为设计，主要完成业务范围内的功能，并通过使用 REST 协议对外提供接口服务。

另一方面，在垂直方向上，主要有 REST API 微服务和 UI 微服务。REST API 微服务提供统一的请求方式来使用业务功能，并通过 Client API 模块对外提供统一的请求接口。REST API 微服务的核心功能通过领域模块来实现，该模块按照分层架构的理念进行设计，可实现数据访问层和业务逻辑层，并以 REST API 微服务为基础实现前后端分离设计，创建 UI 微服务。UI 微服务不直接访问数据，它只专注于人机交互界面的设计，可以通过 Client API 模块来完成。UI 微服务可以是 Web 微服务和 WAP 微服务，分别对应面向 PC 端的应用和面向移动端的应用。

经过两次微服务的划分，再结合高性能和高并发的设计，并通过微服务的多副本发布，就可以构成一个能够适应任何访问规模的、多维的、稳定牢固的网格结构，并且这个网格结构具有自由伸缩的特性，可以根据业务的发展规模进行缩减或扩展，从而可以非常容易地搭建一个可持续扩展的系统平台。

2. 数据库划分方法

在微服务架构中，修改数据库的影响较为广泛，需要保证这种修改是向后兼容的，因此，数据库设计是微服务设计的一个关键点。在微服务架构中，使用单一数据库会存在如下问题。

- 微服务可以提供多个类型的服务，但单一数据库的传统设计会产生紧密耦合，无法进行独立部署的服务。如果有多个服务访问同一数据库，则需要在所有服务之间协调数据模式的更改。在现实工作中，这会导致额外的工作，延迟部署更新。
- 使用单一数据库很难对单个服务进行扩展，这是因为用户只能选择扩展单一数据库。
- 使用单一数据库使得提高应用程序的性能成为挑战。当使用单一数据库时，在一段时间后，最终会形成一个数据庞大的表，使数据检索变得很困难。

在微服务架构中，基本原则是每个微服务都有自己单独的数据库，而且只有微服务本身可以访问这个数据库。微服务之间的数据共享可以通过服务调用或主从表的方式实现。在共享数据时，还需要找到合适的同步方式。

需要注意的是，在微服务架构设计的开始阶段就应该完成数据库划分，使每个微服务模块对应独立的数据库，而且微服务架构一定不是简单的应用组件划分，还包括了后端的数据库划分，这样才是完整意义上的微服务架构。同时，被划分后的数据库之间不应该有交互和集成，所有的交互和集成都应该通过应用组件层的 API 接口服务进行，只有这样才是完整意义上的微服务架构。

6.2.2 微服务项目结构

首先按照水平划分法划分微服务，即从业务维度进行划分，将功能相对独立的业务单独划分为一个微服务，然后将同一业务类型的微服务放在同一项目工程中。网上人才中心系统共划分为 6 个微服务项目，如表 6-1 所示。用户微服务项目主要实现用户管理、角色管理及权限管理；企业微服务项目主要实现企业管理；人才微服务项目主要实现人才个人信息的维护及管理；职位微服务项目主要实现工作职位的发布及管理、应聘的管理；Web 微服务项目主要实现统一的用户界面；基础微服务项目主要实现配置管理、统一认证、网关、注册服务等功能。

表 6-1 网上人才中心系统微服务项目列表

序 号	项 目 名 称	功 能 说 明	数 据 库
1	user-microservice	用户微服务项目	userdb
2	company-microservice	企业微服务项目	companydb
3	person-microservice	人才微服务项目	persondb
4	job-microservice	职位微服务项目	jobdb
5	web-microservice	Web 微服务项目	
6	base-microservice	基础微服务项目	basedb

同一业务类型的微服务包括 REST API 微服务和 UI 微服务，其中，REST API 微服务包含了业务领域的数据建模和接口服务等方面的设计；UI 微服务包含了面向 PC 端的 Web 应用和面向移动端的 WAP 应用等方面的设计。因此，当使用这种组建方法，并根据不同功

能进行模块划分时，一个微服务项目工程大致包含如表 6-2 所示的模块结构。

表 6-2　微服务项目工程的模块结构列表

序　号	模块名称	功能说明	类　型	备　注
1	*-domain	业务领域模块	程序包（jar 包）	实体类、数据访问、业务逻辑
2	*-object	查询对象	程序包（jar 包）	用于查询参数
3	*-restapi	RESTful 接口	微服务应用（war 包）	
4	*-clientapi	客户端接口	程序包（jar 包）	由 Web 应用使用或跨业务微服务使用
5	*-web	PC 端 Web 应用	微服务应用（war 包）	
6	*-wap	移动端 WAP 应用	微服务应用（war 包）	

6.2.3　创建微服务项目

1．创建项目

我们将使用多模块的方式来组建项目。这里以用户微服务项目 user-microservice 为例来说明如何创建项目，基本步骤如下所述。

（1）打开 IDEA，在新建项目窗口的左侧列表框中选择【Maven】选项，如图 6-4 所示，创建一个新项目，然后单击【Next】按钮。

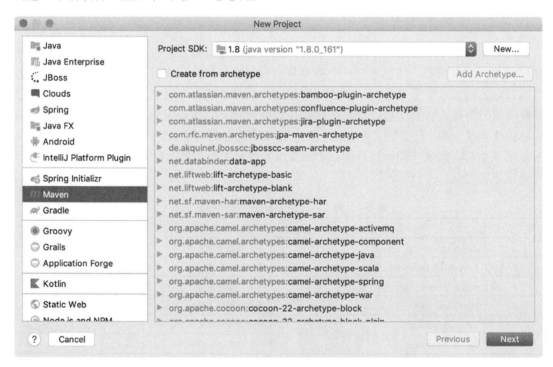

图 6-4　创建一个新项目

（2）输入项目名称，选择项目存放位置，并设置【GroupId】、【ArtifactId】和【Version】，如图 6-5 所示，然后单击【Finish】按钮。

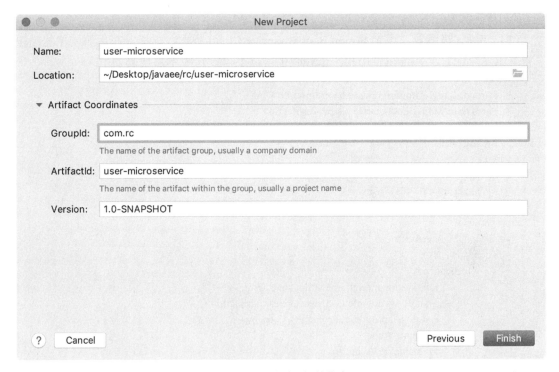

图 6-5　设置项目相关信息

（3）待 IDEA 构建好项目后，项目结构如图 6-6 所示。

图 6-6　项目结构

由于我们将使用多模块的方式来组建项目，因此图 6-6 中生成的 src 目录没什么用，可以将其删除。

2. 基本配置

（1）使用父级配置，设定 Spring Boot 和 Spring Cloud 版本，这样，在后面创建的模块中将继承这个配置，并且在引用相关依赖时也可以不再设定版本，代码如下：

```xml
<parent>
    <groupId>org.springframework.boot</groupId>
    <artifactId>spring-boot-starter-parent</artifactId>
    <version>2.2.6.RELEASE</version>
    <relativePath/>
</parent>
<properties>
    <java.version>1.8</java.version>
    <spring-cloud.version>Hoxton.SR3</spring-cloud.version>
</properties>
<dependencyManagement>
    <dependencies>
        <dependency>
            <groupId>org.springframework.cloud</groupId>
            <artifactId>spring-cloud-dependencies</artifactId>
            <version>${spring-cloud.version}</version>
            <type>pom</type>
            <scope>import</scope>
        </dependency>
    </dependencies>
</dependencyManagement>
```

上述代码中同时定义了 JDK 版本。Spring Boot 使用的版本是 2.2.6.RELEASE，Spring Cloud 使用的版本是 Hoxton.SR3。

（2）引入一些通用配置，以便为所有的模块定义默认的依赖引用。这些依赖引用包含了一些 Spring Boot 开发框架的基本组件，如热部署、自动配置和单元测试等，代码如下：

```xml
<dependency>
    <groupId>org.springframework.boot</groupId>
    <artifactId>spring-boot-devtools</artifactId>
    <scope>runtime</scope>
    <optional>true</optional>
</dependency>
<dependency>
    <groupId>org.springframework.boot</groupId>
    <artifactId>spring-boot-configuration-processor</artifactId>
    <optional>true</optional>
</dependency>
<dependency>
    <groupId>org.springframework.boot</groupId>
    <artifactId>spring-boot-starter-test</artifactId>
    <scope>test</scope>
</dependency>
```

（3）使用统一的插件配置，如果模块中没有相关的插件配置，则默认使用如下配置：

```xml
<build>
    <plugins>
        <plugin>
            <groupId>org.apache.maven.plugins</groupId>
            <artifactId>maven-compiler-plugin</artifactId>
            <version>3.1</version>
            <configuration>
                <source>${java.version}</source>
                <target>${java.version}</target>
            </configuration>
        </plugin>
        <plugin>
            <groupId>org.apache.maven.plugins</groupId>
            <artifactId>maven-surefire-plugin</artifactId>
            <version>2.19.1</version>
            <configuration>
                <skipTests>true</skipTests><!--默认关掉单元测试 -->
            </configuration>
        </plugin>
    </plugins>
</build>
```

整体的配置文件如下：

```xml
<?xml version="1.0" encoding="UTF-8"?>
<project
    xmlns="http://maven.apache.org/POM/4.0.0"
    xmlns:xsi="http://www.w3.org/2001/XMLSchema-instance"
    xsi:schemaLocation="http://maven.apache.org/POM/4.0.0
    https://maven.apache.org/xsd/maven-4.0.0.xsd">
    <modelVersion>4.0.0</modelVersion>
    <parent>
        <groupId>org.springframework.boot</groupId>
        <artifactId>spring-boot-starter-parent</artifactId>
        <version>2.2.6.RELEASE</version>
        <relativePath/>
    </parent>
    <groupId>com.rc</groupId>
    <artifactId>user-microservice</artifactId>
    <version>0.1-SNAPSHOT</version>
    <packaging>pom</packaging>
    <name>user-microservice</name>
    <description>用户微服务</description>
    <properties>
        <java.version>1.8</java.version>
        <spring-cloud.version>Hoxton.SR3</spring-cloud.version>
```

```xml
</properties>
<dependencies>
    <dependency>
        <groupId>org.springframework.boot</groupId>
        <artifactId>spring-boot-devtools</artifactId>
        <scope>runtime</scope>
        <optional>true</optional>
    </dependency>
    <dependency>
        <groupId>org.springframework.boot</groupId>
        <artifactId>spring-boot-configuration-processor</artifactId>
        <optional>true</optional>
    </dependency>
    <dependency>
        <groupId>org.springframework.boot</groupId>
        <artifactId>spring-boot-starter-test</artifactId>
        <scope>test</scope>
    </dependency>
</dependencies>
<dependencyManagement>
    <dependencies>
        <dependency>
            <groupId>org.springframework.cloud</groupId>
            <artifactId>spring-cloud-dependencies</artifactId>
            <version>${spring-cloud.version}</version>
            <type>pom</type>
            <scope>import</scope>
        </dependency>
    </dependencies>
</dependencyManagement>
<build>
    <plugins>
        <plugin>
            <groupId>org.apache.maven.plugins</groupId>
            <artifactId>maven-compiler-plugin</artifactId>
            <version>3.1</version>
            <configuration>
                <source>${java.version}</source>
                <target>${java.version}</target>
            </configuration>
        </plugin>
        <plugin>
            <groupId>org.apache.maven.plugins</groupId>
            <artifactId>maven-surefire-plugin</artifactId>
            <version>2.19.1</version>
```

```xml
                    <configuration>
                        <skipTests>true</skipTests><!--默认关掉单元测试 -->
                    </configuration>
                </plugin>
            </plugins>
        </build>
    </project>
```

6.2.4 创建模块

以之前创建的用户微服务项目 user-microservice 为例来创建模块。

（1）在项目名称上单击鼠标右键，在弹出的快捷菜单中选择【new】→【model】命令，打开【New Module】对话框，如图 6-7 所示。在【New Module】对话框中，选择【Maven】选项，然后单击【Next】按钮。

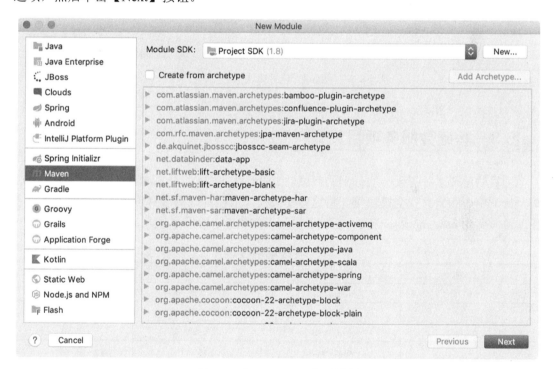

图 6-7 【New Module】对话框

（2）在如图 6-8 所示的对话框中设置模块信息，保持父项目的默认设置为【user-microservice】，输入模块名称为【user-domain】，并在选择模块的存储位置后，单击【Finish】按钮，即可创建模块。

（3）类似地，可创建 user-object、user-restapi、user-clientapi、user-web、user-wap 模块。

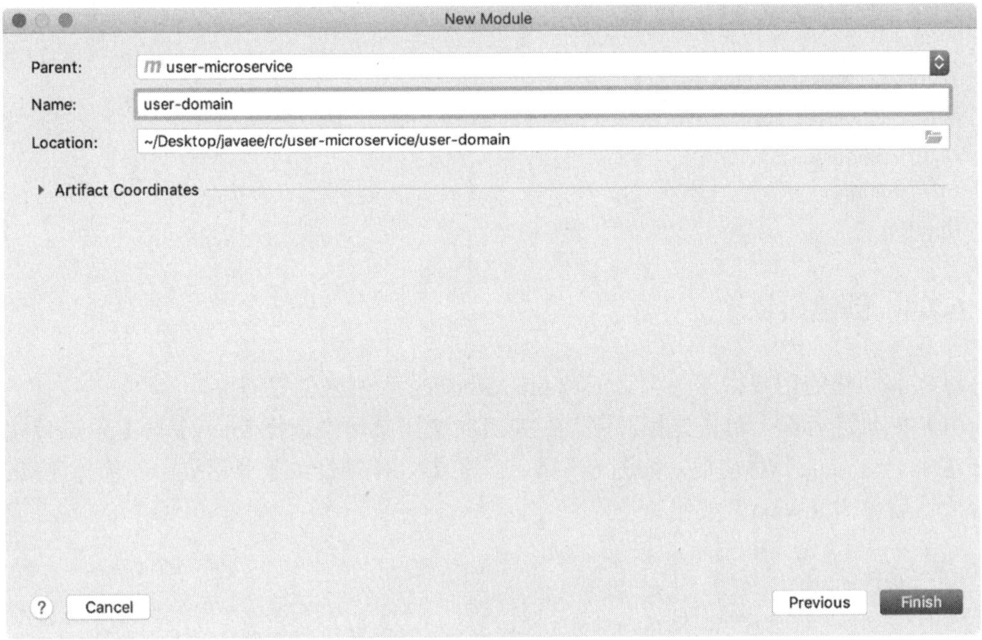

图 6-8　设置模块信息

6.3　基础微服务项目设计

项目的整体运行需要一些基础的环境，如服务注册中心、配置管理中心、微服务网关、统一认证中心和公共安全模块等。将这些基础模块放在基础微服务项目中，并将基础微服务项目命名为 base-microservice。基础模块列表如表 6-3 所示。

表 6-3　基础模块列表

序　号	模　　块	功　　能	类　型
1	base-eureka	服务注册中心	微服务
2	base-config	配置管理中心	微服务
3	base-zuul	微服务网关	微服务
4	base-oauth	统一认证中心	微服务
5	base-security	公共安全模块	程序包

基础微服务项目的创建类似于用户微服务项目 user-microservice。下面介绍服务注册中心、配置管理中心和微服务网关 3 个模块的创建。

6.3.1　创建服务注册中心

服务注册与发现主要基于 Eureka。Eureka 是一个基于 REST 的服务注册与发现的组件。它主要包括以下两部分。

- Eureka Client：一个 Java 客户端，用于简化与 Eureka Server 的交互（通常是微服务中的客户端和服务器端）。
- Eureka Server：提供服务注册与发现的能力（通常是微服务中的服务注册中心）。

创建步骤如下所述。

（1）在 base-microservice 项目中创建 base-eureka 模块。

（2）在 pom 文件中引入如下依赖：

```xml
<dependency>
    <groupId>org.springframework.boot</groupId>
    <artifactId>spring-boot-starter-web</artifactId>
</dependency>
<dependency>
    <groupId>org.springframework.cloud</groupId>
    <artifactId>spring-cloud-starter-netflix-eureka-server</artifactId>
</dependency>
```

（3）在 base-eureka 模块中创建一个启动程序，并将其命名为 EurekaApplication，代码如下：

```java
package com.rc.eureka;
import org.springframework.boot.SpringApplication;
import org.springframework.boot.autoconfigure.SpringBootApplication;
import org.springframework.cloud.netflix.eureka.server.EnableEurekaServer;
@EnableEurekaServer
@SpringBootApplication
public class EurekaApplication {
    public static void main(String[] args) {
        SpringApplication.run(EurekaApplication.class, args);
    }
}
```

（4）在模块的配置文件 application.properties 中增加如下配置：

```
spring.application.name=eureka
server.port=8099
eureka.instance.hostname=localhost
#表示是否将自己注册到 Eureka Server，默认为 true
eureka.client.registerWithEureka=false
#表示是否从 Eureka Server 获取注册信息，默认为 true
eureka.client.fetchRegistry=false
#设置与 Eureka Server 交互的地址，客户端的查询服务和注册服务都需要依赖这个地址
eureka.client.serviceUrl.defaultZone=http://localhost:${server.port}/eureka/
```

6.3.2 创建配置管理中心

配置管理中心可以为所有微服务提供一个统一的配置管理服务。微服务可以使用本地项目的配置，也可以使用配置管理中心的配置，当这两种配置相同时，系统默认优先使用配置管理中心提供的配置。Spring-Cloud-Config 提供了对多种环境的配置文件的支持，如开发环境、测试环境、生产环境等的配置文件。配置文件的命名方式可以采用*-

dev.properties、*-test.properties、*-pro.properties 形式。配置文件可以存放在本地，也可以存放在远程 Git 仓库或远程 SVN 仓库。

1. 创建配置管理中心模块

在 base-microservice 项目中创建一个模块，并将其命名为 base-config，然后在其 pom 文件中引入如下依赖：

```xml
<dependency>
    <groupId>org.springframework.cloud</groupId>
    <artifactId>spring-cloud-starter-netflix-eureka-client</artifactId>
</dependency>
<dependency>
    <groupId>org.springframework.cloud</groupId>
    <artifactId>spring-cloud-config-server</artifactId>
</dependency>
```

eureka-client 是服务注册与发现的工具组件，可以用来提供服务注册中心客户端的功能。
config-server 是配置管理服务器，可以用来创建配置管理中心。

2. 创建一个启动程序

启动程序代码如下：

```java
package com.rc.config;
import org.springframework.boot.SpringApplication;
import org.springframework.boot.autoconfigure.SpringBootApplication;
import org.springframework.cloud.client.discovery.EnableDiscoveryClient;
import org.springframework.cloud.config.server.EnableConfigServer;
@EnableConfigServer
@EnableDiscoveryClient
@SpringBootApplication
public class ConfigApplication {
    public static void main(String[] args) {
        SpringApplication.run(ConfigApplication.class, args);
    }
}
```

3. 在配置文件 bootstrap.properties 中进行配置

这里采用本地文件配置方式，代码如下：

```
spring.application.name=config-service
server.port=9002
#eureka.instance.prefer-ip-address=true
eureka.client.serviceUrl.defaultZone=http://localhost:8099/eureka/
spring.cloud.config.server.native.search-locations=classpath:/config
spring.profiles.active=native
spring.security.user.password=config
```

如果要使用远程 Git 仓库进行配置，则可以进行如下修改：

```
spring.profiles.active=git  spring.cloud.config.server.git.uri= git.uri 指定地址
spring.cloud.config.server.git.default-label= git 仓库 default-label
```

4. 在微服务中使用配置管理中心

以微服务 user-restapi 为例,首先,在应用模块中引入如下依赖:

```xml
<dependency>
    <groupId>org.springframework.cloud</groupId>
    <artifactId>spring-cloud-starter-config</artifactId>
</dependency>
```

然后,在模块中增加配置文件 bootstrap.properties,配置内容类似如下:

```
spring.application.name=user-restapi
eureka.client.serviceUrl.defaultZone=http://localhost:8099/eureka/
server.port=8889
#对应{profile},指定客户端当前的环境,可选值包括 dev、test、pro
spring.cloud.config.profile=dev
#开启配置信息发现
spring.cloud.config.discovery.enabled=true
#指定配置中心的 service-id,便于扩展为高可用配置集群
spring.cloud.config.discovery.service-id=base-config
spring.cloud.config.fail-fast=true
spring.cloud.config.username=user
spring.cloud.config.password=config
spring.main.allow-bean-definition-overriding=true
```

5. 在 resources/config 文件下创建其他微服务配置

例如,对于微服务 user-restapi 而言,可以将开发模式的配置文件命名为 user-restapi-dev.properties,内容如下:

```
#配置 Tomcat 编码,默认为 UTF-8
server.tomcat.uri-encoding=UTF-8
#配置最大线程数
server.tomcat.max-threads=1000
spring.datasource.url=jdbc:mysql://localhost:3306/rc?useUnicode=true&characterEncoding=utf8&serverTimezone=GMT&useSSL=false&zeroDateTimeBehavior=convertToNull&allowMultiQueries=true
spring.datasource.username=root
spring.datasource.password=12345
spring.datasource.driverClassName=com.mysql.cj.jdbc.Driver
spring.datasource.type=com.alibaba.druid.pool.DruidDataSource

#连接池的配置信息
#初始化时建立物理连接的个数
spring.datasource.druid.initial-size=3
#连接池最小连接数
spring.datasource.druid.min-idle=3
#连接池最大连接数
spring.datasource.druid.max-active=20
#获取连接时最大等待时间,单位为毫秒
spring.datasource.druid.max-wait=60000
#申请连接时检测,若空闲时间大于 timeBetweenEvictionRunsMillis,则执行 validationQuery,检测
#连接是否有效
spring.datasource.druid.test-while-idle=true
#既作为检测的间隔时间,又作为 testWhileIdel 执行的依据
```

```
spring.datasource.druid.time-between-connect-error-millis=60000
#销毁线程时检测，如果当前连接的最后活动时间和当前时间差大于该值，则关闭当前连接
spring.datasource.druid.min-evictable-idle-time-millis: 30000
#用来检测连接是否有效的 SQL 语句必须是查询语句
#MySQL 中为 select 'x'
#Oracle 中为 select 1 from dual
spring.datasource.druid.validation-query=select 'x'
#申请连接时会执行 validationQuery，检测连接是否有效，若开启，则会降低性能，默认为 true
spring.datasource.druid.test-on-borrow=false
#归还连接时会执行 validationQuery，检测连接是否有效，若开启，则会降低性能，默认为 true
spring.datasource.druid.test-on-return=false
#是否缓存 preparedStatement，MySQL 5.5+建议开启
spring.datasource.druid.pool-prepared-statements=true
#当值大于 0 时，poolPreparedStatements 会自动修改为 true
spring.datasource.druid.max-pool-prepared-statement-per-connection-size=20
#合并多个 DruidDataSource 的监控数据
spring.datasource.druid.use-global-data-source-stat=false
#配置扩展插件
#spring.datasource.druid.filters=stat,wall,slf4j
#通过 connect-properties 属性来打开 mergeSql 功能，进行慢 SQL 记录
spring.datasource.druid.connect-properties = druid.stat.mergeSql=true; druid.stat.slowSqlMillis=5000
#定时输出统计信息到日志中，并且每次输出日志会导致连接池相关的计数器清零（reset）
spring.datasource.druid.time-between-log-stats-millis=300000
#配置 DruidStatFilter
spring.datasource.druid.web-stat-filter.enabled=true
spring.datasource.druid.web-stat-filter.url-pattern=/*
spring.datasource.druid.web-stat-filter.exclusions=*.js,*.gif,*.jpg,*.bmp,*.png,*.css,*.ico,/druid/*
#配置 DruidStatViewServlet
#是否启用 StatViewServlet（监控页面），默认值为 false（考虑到安全问题，默认不启动，
#如果需要启用，建议设置密码或白名单以保障安全）
spring.datasource.druid.stat-view-servlet.enabled=true
spring.datasource.druid.stat-view-servlet.url-pattern=/druid/*
#IP 白名单（如果没有配置或为空，则允许所有访问）
spring.datasource.druid.stat-view-servlet.allow=127.0.0.1,192.168.0.1
#IP 黑名单（当共同存在时，deny 优先于 allow）
spring.datasource.druid.stat-view-servlet.deny=192.168.0.128
#禁用 HTML 页面上的 Reset All 功能
spring.datasource.druid.stat-view-servlet.reset-enable=false
#登录名
spring.datasource.druid.stat-view-servlet.login-username=admin
#登录密码
spring.datasource.druid.stat-view-servlet.login-password=123456

#MyBatis-Plus 相关配置
#XML 扫描，多个目录用逗号或分号分隔（告诉 Mapper 所对应的 XML 文件位置）
mybatis-plus.mapper-locations=classpath:mapper/*.xml
#以下配置均有默认值，可以不设置
#主键类型。AUTO 表示数据库 ID 自增；INPUT 表示用户输入 ID；ID_WORKER 表示全局唯一 ID
#（数字类型唯一 ID）；UUID 表示全局唯一 ID
```

```
mybatis-plus.global-config.db-config.id-type=AUTO
#字段策略。IGNORED 表示忽略判断；NOT_NULL 表示非 NULL 判断；NOT_EMPTY 表示非空判断
mybatis-plus.global-config.db-config.field-strategy=NOT_EMPTY
#数据库类型
mybatis-plus.global-config.db-config.db-type=MYSQL
#是否开启自动驼峰命名规则映射：从数据库列名到 Java 属性驼峰命名的类似映射
mybatis-plus.configuration.map-underscore-to-camel-case=true
#在返回 Map 时，true 表示当查询数据为空时，字段返回为 null，false 表示当查询数据为空时，
#字段将会被隐藏
mybatis-plus.configuration.call-setters-on-nulls=true
#这个配置会将执行的 SQL 语句打印出来，在开发或测试时可以使用
mybatis-plus.configuration.log-impl=org.apache.ibatis.logging.stdout.StdOutImpl
#扫描位置
mybatis-plus.typeAliasesPackage=com.rc.user.entity
#逻辑删除指定值
mybatis-plus.global-config.db-config.logic-delete-value=1
mybatis-plus.global-config.db-config.logic-not-delete-value=
spring.aop.proxy-target-class=true
#security.oauth2.resource.id=person-restapi
#security.oauth2.client.user-authorization-uri=http://localhost:5777/oauth-server/user

ribbon.eureka.enabled=true
feign.client.config.default.connect-timeout=60000
feign.client.config.default.read-timeout=60000
```

6.3.3 创建微服务网关

不同的微服务一般会有不同的网络地址，而客户端可能需要调用多个服务接口才能完成一个业务需求。如果让客户端直接与各微服务通信，则会产生如下问题。
- 客户端会多次请求不同的微服务，增加了客户端的复杂性。
- 存在跨域请求，处理相对复杂。
- 认证复杂，每个服务都需要进行独立认证。
- 难以重构，多个服务可能会合并为一个或划分为多个。

微服务网关位于服务端与客户端的中间层，所有外部服务请求都会先经过微服务网关，客户只能与微服务网关进行交互，无须调用特定微服务接口，使得开发得到简化。微服务网关还具备如下优点。
- 易于监控：微服务网关可以收集监控数据并进行分析。
- 易于认证：可以在微服务网关上进行认证，然后将请求转发给微服务，无须对每个微服务都进行认证。
- 减少客户端与微服务之间的交互次数。

Zuul 是开源的微服务网关，可以与 Eureka、Ribbon、Hystrix 等组件配合使用。Zuul 的核心是一系列过滤器，这些过滤器可以完成以下功能。
- 身份认证与安全：识别每个资源的验证要求，并拒绝不符合要求的请求。

- 审核与监控：在边缘位置追踪有意义的数据和统计结果，从而得到精确的生产视图。
- 动态路由：动态地将请求路由到不同的后端集群。
- 压力测试：逐渐增加指向集群的流量，以了解性能。
- 负载分配：为每种负载类型分配对应容量，并弃用超出限定值的请求。
- 静态响应处理：在边缘位置直接创建部分响应，从而避免转发到内部集群。
- 多区域弹性：跨越 AWS Region 进行请求路由，实现 ELB 使用的多样化，让系统边缘更贴近使用者。

在网上人才中心系统中，网关主要用于路由，负责均衡及权限控制。

1. 创建微服务网关模块

在 base-microservice 项目中创建一个模块，并将其命名为 base-zuul，然后在其 pom 文件中引入如下依赖：

```xml
<dependency>
    <groupId>org.springframework.cloud</groupId>
    <artifactId>spring-cloud-starter-netflix-eureka-client</artifactId>
</dependency>
<dependency>
    <groupId>org.springframework.cloud</groupId>
    <artifactId>spring-cloud-starter-netflix-zuul</artifactId>
</dependency>
<dependency>
    <groupId>org.springframework.cloud</groupId>
    <artifactId>spring-cloud-starter-netflix-ribbon</artifactId>
</dependency>
<dependency>
    <groupId>org.springframework.cloud</groupId>
    <artifactId>spring-cloud-starter-netflix-hystrix</artifactId>
</dependency>
```

2. 创建一个启动程序

启动程序代码如下：

```java
package com.rc.zuul;
import org.springframework.boot.SpringApplication;
import org.springframework.boot.autoconfigure.SpringBootApplication;
import org.springframework.cloud.client.discovery.EnableDiscoveryClient;
import org.springframework.cloud.netflix.zuul.EnableZuulProxy;
@SpringBootApplication
@EnableZuulProxy
@EnableDiscoveryClient
public class ZuulApplication {
    public static void main(String[] args) {
        SpringApplication.run(ZuulApplication.class, args);
    }
}
```

3. 配置文件 bootstrap.properties

```
spring.application.name=gateway-server
eureka.client.service-url.defaultZone=http://localhost:8099/eureka/
server.port=8084


zuul.routes.eureka-client1.path=/person-web/**
zuul.routes.eureka-client1.sensitiveHeaders=*
zuul.routes.eureka-client1.serviceId=person-web
zuul.routes.eureka-client1.strip-prefix=false

zuul.routes.eureka-client2.path=/company-web/**
zuul.routes.eureka-client2.sensitiveHeaders=*
zuul.routes.eureka-client2.serviceId=company-web
zuul.routes.eureka-client2.strip-prefix=false

zuul.routes.eureka-client3.path=/job-web/**
zuul.routes.eureka-client3.sensitiveHeaders=*
zuul.routes.eureka-client3.serviceId=job-web
zuul.routes.eureka-client3.strip-prefix=false

zuul.routes.eureka-client4.path=/user-web/**
zuul.routes.eureka-client4.sensitiveHeaders=*
zuul.routes.eureka-client4.serviceId=user-web
zuul.routes.eureka-client4.strip-prefix=false

zuul.routes.eureka-client5.path=/web/**
zuul.routes.eureka-client5.sensitiveHeaders=*
zuul.routes.eureka-client5.serviceId=web-client
zuul.routes.eureka-client5.strip-prefix=false

zuul.add-proxy-headers=true
#取消重试机制
zuul.retryable=false
#host-header
zuul.add-host-header=true
#请求处理超时时间
ribbon.ReadTimeout=5000
#请求连接超时时间
ribbon.ConnectTimeout=1000
#断路器超时时间
hystrix.command.default.execution.isolation.thread.timeoutInMilliseconds=900
```

6.4 REST API 微服务设计

6.4.1 领域业务设计

领域业务设计是基于面向对象思想的领域驱动设计（Domain Driven Design，DDD）。

领域业务模块设计主要涉及实体类、数据访问层和业务逻辑层设计。数据访问层的实现仍然采用 MyBatis-Plus。

下面以用户微服务项目 user-microservice 为例进行说明。

1．创建 user-domain 模块

在用户微服务项目 user-microservice 中创建一个模块，并将其命名为 user-domain，然后在其 pom 文件中引入如下依赖：

```xml
<dependency>
        <groupId>org.apache.commons</groupId>
        <artifactId>commons-lang3</artifactId>
        <version>3.10</version>
</dependency>
<dependency>
        <groupId>com.baomidou</groupId>
        <artifactId>mybatis-plus-boot-starter</artifactId>
        <version>3.3.1</version>
</dependency>
```

2．在模块中创建包

包的结构如图 6-9 所示。

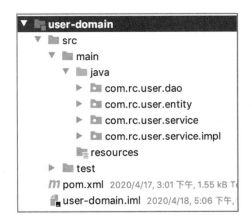

图 6-9　包的结构

3．在 entity 包下创建实体类

创建 3 个实体类。与前面章节所不同的是，用户与企业的关联关系需要取消。按照微服务架构设计，用户表和企业表可被存放在不同的数据库中。为了提高性能，可以不定义跨数据库关联，直接在数据表中存储企业号和企业名称。同时，在实体类中可直接定义 comId 和 comName 属性。

但是，用户表、角色表、权限表等相关表会被存放在一个数据库中，因此仍然需要定义关联关系。

1）用户实体类

```
package com.rc.user.entity;
import com.baomidou.mybatisplus.annotation.IdType;
```

```java
import com.baomidou.mybatisplus.annotation.TableName;
import com.baomidou.mybatisplus.extension.activerecord.Model;
import com.baomidou.mybatisplus.annotation.TableId;
import java.io.Serializable;
import java.util.List;
import io.swagger.annotations.ApiModel;
import lombok.Data;
import lombok.EqualsAndHashCode;
import lombok.ToString;
import lombok.experimental.Accessors;
@Data
@EqualsAndHashCode(callSuper = false)
@Accessors(chain = true)
@TableName(value = "rc_user", resultMap = "BaseResultMap")
public class RcUser extends Model<RcUser>{
    @TableId(value = "user_id", type = IdType.AUTO)
    private Integer userId;
    private String userName;
    private String userRealname;
    private String userPwd;
    private String userType;
    private String userEmail;
    private String userPosttime;
    private Integer comId;
    private String comName;
    @ToString.Exclude
    private List<RcRole> roles;
    @Override
    protected Serializable pkVal() {
        return this.getUserId();
    }
}
```

2）角色实体类

```java
package com.rc.user.entity;
import com.baomidou.mybatisplus.annotation.IdType;
import com.baomidou.mybatisplus.annotation.TableName;
import com.baomidou.mybatisplus.extension.activerecord.Model;
import com.baomidou.mybatisplus.annotation.TableId;
import java.io.Serializable;
import java.util.List;
import lombok.Data;
import lombok.EqualsAndHashCode;
import lombok.ToString;
import lombok.experimental.Accessors;
@Data
@EqualsAndHashCode(callSuper = false)
@Accessors(chain = true)
@TableName(value = "rc_role",resultMap = "BaseResultMap")
public class RcRole extends Model<RcRole> {
```

```java
    @TableId(value = "role_id", type = IdType.AUTO)
    private Integer roleId;
    private String roleName;
    private String roleKey;
    @ToString.Exclude
    private List<RcPermission> permissions;
    @TableField(exist = false)
    private Boolean isSelected;
    @Override
    protected Serializable pkVal() {
        return this.roleId;
    }
}
```

3）权限实体类

```java
package com.rc.user.entity;
import com.baomidou.mybatisplus.annotation.IdType;
import com.baomidou.mybatisplus.annotation.TableField;
import com.baomidou.mybatisplus.annotation.TableName;
import com.baomidou.mybatisplus.extension.activerecord.Model;
import com.baomidou.mybatisplus.annotation.TableId;
import java.io.Serializable;
import java.util.List;
import lombok.Data;
import lombok.EqualsAndHashCode;
import lombok.experimental.Accessors;
@Data
@EqualsAndHashCode(callSuper = false)
@Accessors(chain = true)
@TableName(value = "rc_permission",resultMap = "BaseResultMap")
public class RcPermission extends Model<RcPermission> {
    @TableId(value = "perm_id", type = IdType.AUTO)
    private Integer permId;
    private String permName;
    private String permUrl;
    private List<RcRole> roles;
    @TableField(exist = false)
    private Boolean isSelected;
    @Override
    protected Serializable pkVal() {
        return this.permId;
    }
}
```

4．数据访问层和业务逻辑层的设计

数据访问层和业务逻辑层的设计基本和前面章节中的一样，这里不再赘述了。

6.4.2 查询对象设计

将查询对象放在一个模块中，可以方便其他模块使用。之所以定义查询对象，主要是为了避免其他模块与数据访问层实现技术耦合，因为在实体类定义时都使用了 ORM 框架的标注。

下面以用户微服务项目 user-microservice 为例进行说明。

在用户微服务项目 user-microservice 中创建一个模块，并将其命名为 user-object。然后创建 com.rc.user.object 包，并在该包下创建以下 3 个类。

1）用户查询对象类

```
package com.rc.user.object;
import lombok.Data;
import java.util.List;
@Data
public class RcUserQo{
    private Integer userId;
    private String userName;
    private String userRealname;
    private String userPwd;
    private String userType;
    private String userEmail;
    private String userPosttime;
    private Integer comId;
    private String comName;
    private List<RcRoleQo> roles;
}
```

2）角色查询对象类

```
package com.rc.user.object;
import lombok.Data;
import java.util.List;
@Data
public class RcRoleQo {
    private Integer roleId;
    private String roleName;
    private String roleKey;
    private List<RcPermissionQo> permissions;
}
```

3）权限查询对象类

```
package com.rc.user.object;
import lombok.Data;
import java.util.List;
@Data
public class RcPermissionQo {
    private Integer permId;
    private String permName;
    private String permUrl;
    private List<RcRoleQo> roles;
}
```

6.4.3　REST API 应用设计

在完成 domain 模块和 object 模块后，就可以进行 REST API 微服务应用设计了。REST API 是独立的 Web 应用，可以独立部署、独立运行，并且使用独立的数据库。REST API 可以根据需要进行多实例部署。

下面以用户微服务项目 user-microservice 为例来说明如何设计 REST API。

1．创建 user-restapi 模块

在用户微服务项目 user-microservice 中创建一个模块，并将其命名为 user-restapi，然后在其 pom 文件中引入如下依赖：

```
<dependency>
    <groupId>org.springframework.boot</groupId>
    <artifactId>spring-boot-starter</artifactId>
</dependency>
<dependency>
    <groupId>org.springframework.boot</groupId>
    <artifactId>spring-boot-starter-web</artifactId>
</dependency>
<dependency>
    <groupId>org.springframework.cloud</groupId>
    <artifactId>spring-cloud-starter-openfeign</artifactId>
</dependency>
<dependency>
    <groupId>org.springframework.cloud</groupId>
    <artifactId>spring-cloud-starter-netflix-eureka-client</artifactId>
</dependency>
<dependency>
    <groupId>org.springframework.boot</groupId>
    <artifactId>spring-boot-starter-tomcat</artifactId>
    <scope>provided</scope>
</dependency>
<dependency>
    <groupId>org.springframework.cloud</groupId>
    <artifactId>spring-cloud-starter-config</artifactId>
</dependency>
<dependency>
    <groupId>org.springframework.boot</groupId>
    <artifactId>spring-boot-starter-actuator</artifactId>
</dependency>
<dependency>
    <groupId>mysql</groupId>
    <artifactId>mysql-connector-java</artifactId>
    <scope>runtime</scope>
</dependency>
<dependency>
    <groupId>com.alibaba</groupId>
    <artifactId>druid-spring-boot-starter</artifactId>
```

```xml
        <version>1.1.20</version>
    </dependency>
    <!-- https://mvnrepository.com/artifact/org.apache.commons/commons-lang3 -->
    <dependency>
        <groupId>org.apache.commons</groupId>
        <artifactId>commons-lang3</artifactId>
        <version>3.10</version>
    </dependency>
    <dependency>
        <groupId>com.rc</groupId>
        <artifactId>user-domain</artifactId>
        <version>0.1-SNAPSHOT</version>
    </dependency>
    <dependency>
        <groupId>com.rc</groupId>
        <artifactId>user-object</artifactId>
        <version>0.1-SNAPSHOT</version>
    </dependency>
```

上述配置中引用了 domain 模块和 object 模块。

2．创建启动程序

```java
package com.rc.user;
import com.alibaba.druid.filter.Filter;
import com.alibaba.druid.pool.DruidDataSource;
import com.alibaba.druid.wall.WallConfig;
import com.alibaba.druid.wall.WallFilter;
import com.baomidou.mybatisplus.extension.plugins.PaginationInterceptor;
import org.springframework.boot.SpringApplication;
import org.springframework.boot.autoconfigure.SpringBootApplication;
import org.springframework.boot.context.properties.ConfigurationProperties;
import org.springframework.cloud.client.discovery.EnableDiscoveryClient;
import org.springframework.cloud.openfeign.EnableFeignClients;
import org.springframework.context.annotation.Bean;
import org.springframework.transaction.annotation.EnableTransactionManagement;
import javax.sql.DataSource;
import java.util.ArrayList;
import java.util.List;
@SpringBootApplication
@EnableDiscoveryClient
@EnableTransactionManagement
@EnableFeignClients
public class UserRestApiApplication {

    public static void main(String[] args) {
        SpringApplication.run(UserRestApiApplication.class, args);
    }

    @Bean
    @ConfigurationProperties(prefix = "spring.datasource")
```

```java
    public DataSource dataSource() {
        DruidDataSource druidDataSource = new DruidDataSource();
        List<Filter> filterList=new ArrayList<>();
        filterList.add(wallFilter());
        druidDataSource.setProxyFilters(filterList);
        return druidDataSource;
    }
    @Bean
    public WallFilter wallFilter(){
        WallFilter wallFilter=new WallFilter();
        wallFilter.setConfig(wallConfig());
        return wallFilter;
    }
    @Bean
    public WallConfig wallConfig(){
        WallConfig config =new WallConfig();
        config.setMultiStatementAllow(true);      //允许一次执行多条语句
        config.setNoneBaseStatementAllow(true);   //允许非基本语句的其他语句
        return config;
    }
    @Bean
    public PaginationInterceptor paginationInterceptor() {
        return new PaginationInterceptor();
    }
}
```

在上述代码中,增加了对 Druid 的配置,以便支持一次执行多条语句。此外,上述代码对 MyBatis-Plus 分页功能进行了配置。

3. 模块的配置文件

在模块中通过 bootstrap.properties 文件进行配置,在配置管理中心中通过 user-restapi-dev.properties 文件进行配置。

4. 设计请求处理控制类

(1) 创建一个基类,以便统一处理异常、统一返回数据格式,代码如下:

```java
package com.rc.user.controller;
//导包语句省略
...
public abstract RcBaseController {
    static class ResultModel extends ModelMap {
        public static final String SUCCESS_CODE = "1";
        public static final String FAILURE_CODE = "0";
        private ResultModel(){
        }
        public static ResultModel of(String code, Object data) {
            ResultModel obj = new ResultModel();
            obj.addAttribute("code", code);
            obj.addAttribute("data", data);
            return obj;
```

```
        }
    }
}
```

（2）创建用户控制类，代码如下：

```java
package com.rc.user.controller;
//导包语句省略
...
@Slf4j
@RestController
@RequestMapping("/user")
public class RcUserController extends RcBaseController {
    @Resource
    public RcUserService rcUserService;
    @Resource
    private RcRoleService rcRoleService;
    public PasswordEncoder passwordEncoder = new BCryptPasswordEncoder();
    @PostMapping("/register")
    @Transactional
    public CompletableFuture<ModelMap> register(@RequestBody RcUserQo rcUser) {
        return CompletableFuture.supplyAsync(() -> {
            try {
                ObjectMapper objectMapper = new ObjectMapper();
                RcUser user=objectMapper.convertValue(rcUser, RcUser.class);
                rcUser.setUserType("个人用户");
                rcUser.setUserPwd(passwordEncoder.encode(rcUser.getUserPwd()));
                rcUserService.save(user);
                return ResultModel.of(ResultModel.SUCCESS_CODE, "注册成功！");
            } catch (Exception e) {
                log.debug(e.getMessage());
                return ResultModel.of(ResultModel.FAILURE_CODE, "注册失败！");
            }
        });
    }
    @PostMapping("/add")
    @Transactional
    public CompletableFuture<ModelMap> add(@RequestBody RcUserQo rcUser){
        return CompletableFuture.supplyAsync(() -> {
            try {
                ObjectMapper objectMapper = new ObjectMapper();
                RcUser user=objectMapper.convertValue(rcUser, RcUser.class);
                rcUserService.addUser(user);
                return ResultModel.of(ResultModel.SUCCESS_CODE, "添加成功！");
            } catch (Exception e) {
                log.debug(e.getMessage());
                return ResultModel.of(ResultModel.FAILURE_CODE, "添加失败！");
            }
        });
    }
    @PostMapping("/personalEdit")
```

```java
@Transactional
public CompletableFuture<ModelMap> personalEdit(@RequestBody RcUserQo rcUser) {
    return CompletableFuture.supplyAsync(() -> {
        try {
            RcUser user1 = rcUserService.getById(rcUser.getUserId());
            user1.setUserName(rcUser.getUserName());
            user1.setUserRealname(rcUser.getUserRealname());
            user1.setUserEmail(rcUser.getUserEmail());
            rcUserService.editUser(user1);
            return ResultModel.of(ResultModel.SUCCESS_CODE, "修改成功！");
        } catch (Exception e) {
            log.debug(e.getMessage());
            return ResultModel.of(ResultModel.FAILURE_CODE, "修改失败！");
        }
    });
}
@PostMapping("/edit")
@Transactional
public CompletableFuture<ModelMap> edit(@RequestBody RcUserQo rcUser) {
    return CompletableFuture.supplyAsync(() -> {
        try {
            ObjectMapper objectMapper = new ObjectMapper();
            RcUser user=objectMapper.convertValue(rcUser, RcUser.class);
            rcUserService.editUser(user);
            return ResultModel.of(ResultModel.SUCCESS_CODE, "修改成功！");
        } catch (Exception e) {
            log.debug(e.getMessage());
            return ResultModel.of(ResultModel.FAILURE_CODE, "修改失败！");
        }
    });
}
@PostMapping("/forFindPwd")
public CompletableFuture<ModelMap> forFindPwd(String userName, String userEmail, String newPwd) {
    return CompletableFuture.supplyAsync(() -> {
        try {
            QueryWrapper<RcUser> queryWrapper = new QueryWrapper<>();
            queryWrapper.eq("user_name", userName);
            queryWrapper.eq("user_email", userEmail);
            RcUser rcUser = rcUserService.getOne(queryWrapper);
            if(rcUser!=null) {
                rcUser.setUserPwd(passwordEncoder.encode(newPwd));
                rcUserService.editUser(rcUser);
                return ResultModel.of(ResultModel.SUCCESS_CODE, "找回成功！");
            }else{
                return ResultModel.of(ResultModel.SUCCESS_CODE, "输入的信息有误！");
            }
        } catch (Exception e) {
            log.debug(e.getMessage());
```

```java
                    return ResultModel.of(ResultModel.FAILURE_CODE, "找回密码失败！");
                }
        });
    }
    @GetMapping("/willEditPwd")
    public CompletableFuture<ModelMap> willEditPwd() {
        return CompletableFuture.supplyAsync(() -> {
            try {
                //RcUser rcUser = this.getCurrentUser();
                //return ResultModel.of(ResultModel.SUCCESS_CODE, rcUser.getUserId());
                return null;
            } catch (Exception e) {
                log.debug(e.getMessage());
                return ResultModel.of(ResultModel.FAILURE_CODE, "修改密码失败！");
            }
        });
    }
    @PostMapping("/editPwd")
    @Transactional
    public CompletableFuture<ModelMap> editPwd(Integer userId, String userPwd, String newPwd) {
        return CompletableFuture.supplyAsync(() -> {
            try {
                RcUser rcUser = rcUserService.getById(userId);
                boolean f = passwordEncoder.matches(userPwd, rcUser.getUserPwd());
                if (f) {
                    rcUser.setUserPwd(passwordEncoder.encode(newPwd));
                    rcUserService.editUser(rcUser);
                    return ResultModel.of(ResultModel.SUCCESS_CODE, rcUser.getUserId());
                } else {
                    return ResultModel.of(ResultModel.FAILURE_CODE, "原密码错误！");
                }
            } catch (Exception e) {
                log.debug(e.getMessage());
                return ResultModel.of(ResultModel.FAILURE_CODE, "修改密码失败！");
            }
        });
    }
    @GetMapping("/delete/{id}")
    @Transactional
    public CompletableFuture<ModelMap> delete(@PathVariable("id") Integer id) {
        return CompletableFuture.supplyAsync(() -> {
            try {
                rcUserService.removeById(id);
                return ResultModel.of(ResultModel.SUCCESS_CODE, "删除成功！");
            } catch (Exception e) {
                log.debug(e.getMessage());
                return ResultModel.of(ResultModel.FAILURE_CODE, "删除失败！");
            }
        });
```

```java
}
@PostMapping("/delete/batch")
@Transactional
public CompletableFuture<ModelMap> deleteBatch(String ids){
    return CompletableFuture.supplyAsync(() -> {
        try {
            List<Integer> idList = new ArrayList<Integer>();
            for (String id : ids.split(",")) {
                idList.add(Integer.valueOf(id));
            }
            rcUserService.removeByIds(idList);
            return ResultModel.of(ResultModel.SUCCESS_CODE, "删除成功！");
        } catch (Exception e) {
            log.debug(e.getMessage());
            return ResultModel.of(ResultModel.FAILURE_CODE, "删除失败！");
        }
    });
}
@GetMapping("/willEditRole/{id}")
public CompletableFuture<ModelMap> willEditRole(@PathVariable("id") Integer id) {
    return CompletableFuture.supplyAsync(() -> {
        try {
            RcUser rcUser = rcUserService.getById(id);
            List<RcRole> list = rcRoleService.list();
            ModelMap map = new ModelMap();
            map.put("userId", id);
            for (RcRole r : list) {
                r.setIsSelected(false);
                for (RcRole r1 : rcUser.getRoles()) {
                    if (r.getRoleId() == r1.getRoleId()) {
                        r.setIsSelected(true);
                        break;
                    }
                }
            }
            map.put("list", list);
            return ResultModel.of(ResultModel.SUCCESS_CODE, map);
        } catch (Exception e) {
            log.debug(e.getMessage());
            return ResultModel.of(ResultModel.FAILURE_CODE, "删除失败！");
        }
    });
}
@PostMapping("/editRole/{id}")
@Transactional
public CompletableFuture<ModelMap> editRole(@PathVariable("id") Integer id, Integer[] roleIds) {
    return CompletableFuture.supplyAsync(() -> {
        try {
            rcUserService.editRole(id, roleIds);
```

```java
                return ResultModel.of(ResultModel.SUCCESS_CODE, "修改角色成功！");
            } catch (Exception e) {
                log.debug(e.getMessage());
                return ResultModel.of(ResultModel.FAILURE_CODE, "修改角色失败！");
            }
        });
    }
    @GetMapping("/find/{id}")
    public CompletableFuture<ModelMap> find(@PathVariable("id") Integer id) {
        return CompletableFuture.supplyAsync(() -> {
            try {
                RcUser rcUser = rcUserService.getById(id);
                return ResultModel.of(ResultModel.SUCCESS_CODE, rcUser);
            } catch (Exception e) {
                log.debug(e.getMessage());
                return ResultModel.of(ResultModel.FAILURE_CODE, "按ID查询失败！");
            }
        });
    }
    @RequestMapping("/findByUserName")
    public CompletableFuture<ModelMap> findByUserName(String userName) {
        return CompletableFuture.supplyAsync(() -> {
            try {
                RcUser rcUser = rcUserService.findUser(userName);
                return ResultModel.of(ResultModel.SUCCESS_CODE, rcUser);
            } catch (Exception e) {
                log.debug(e.getMessage());
                return ResultModel.of(ResultModel.FAILURE_CODE, "按用户名查询失败！");
            }
        });
    }
    @RequestMapping("/manage")
    public CompletableFuture<ModelMap> manage(String userName, final Integer pageNo, final Integer pageSize) {
        return CompletableFuture.supplyAsync(() -> {
            try {
                ModelMap map = new ModelMap();
                QueryWrapper<RcUser> queryWrapper = new QueryWrapper<>();
                if (StringUtils.isNotBlank(userName)) {
                    queryWrapper.like("user_name", userName);
                }
                queryWrapper.orderByDesc("user_posttime");
                IPage<RcUser> page = rcUserService.page(new Page<>(pageNo, pageSize), queryWrapper);
                map.put("userName", userName);
                map.put("total", page.getTotal());
                map.put("pageNo", pageNo);
                map.put("pageSize", pageSize);
                map.put("rows", page.getRecords());
```

```
                    return ResultModel.of(ResultModel.SUCCESS_CODE, map);
                } catch (Exception e) {
                    log.debug(e.getMessage());
                    return ResultModel.of(ResultModel.FAILURE_CODE, "分页查询失败！");
                }
            });
    }
}
```

在上述代码中，所有方法统一返回 ResultModel 类型数据，该类型继承于 ModelMap。ResultModel 不允许通过 new 创建对象，只能通过 of()方法创建对象，以确保返回的数据具有统一的格式，即包含 code 和 data。code 表示请求是否成功，其值为 SUCCESS_CODE 表示成功；其值为 FAILURE_CODE 表示失败。data 用于存放返回的数据，若是增加、修改、删除操作，则返回是否成功的信息；若是查询操作，则查询成功会返回查询的结果，查询不成功会返回查询失败信息。

此外，为了提高性能，适应高并发性，此次编程使用了 Java 非阻塞异步编程方法 CompletableFuture。

6.4.4 RESTful 的 HTTP 接口设计

针对 REST API 微服务对外提供 RESTful 的 HTTP 接口，我们使用 Spring Cloud 组件中的 FeignClient 来实现。这个组件与 RestTemplate 相比更加高效和实用，而且使用方法更加简单。使用 FeignClient 通过声明式接口就可以实现 HTTP 调用，并且在调用的过程中，将自动使用 Zuul 的动态路由和 Ribbon 的负载均衡服务。

我们将接口存放在一个模块中，以便打包并被其他微服务引用。

1. 创建 user-clientapi 模块

在用户微服务项目 user-microservice 中创建一个模块，并将其命名为 user-clientapi，然后在其 pom 文件中引入如下依赖：

```
<dependency>
    <groupId>org.springframework.cloud</groupId>
    <artifactId>spring-cloud-starter-openfeign</artifactId>
</dependency>
<dependency>
    <groupId>org.springframework.cloud</groupId>
    <artifactId>spring-cloud-starter-netflix-hystrix</artifactId>
</dependency>
<dependency>
    <groupId>com.rc</groupId>
    <artifactId>user-object</artifactId>
    <version>0.1-SNAPSHOT</version>
</dependency>
```

2. 定义 FeignClient 接口

在下面的接口定义中，使用@FeignClient 标注设定了所引用的 REST API 应用。同时，使用了自定义的熔断处理类 RcPermissionClientApiFallback，代码如下：

```java
package com.rc.user.clientapi;
//导包语句省略
...
@Component
@FeignClient(value = "user-restapi",fallbackFactory = RcPermissionClientApiFallback.class)
public interface RcPermissionClientApi {
    @PostMapping("/permission/add")
    public ModelMap add(@RequestBody RcPermissionQo rcPermission) ;
    @PostMapping("/permission/edit")
    public ModelMap edit(@RequestBody RcPermissionQo rcPermission);
    @GetMapping("/permission/delete/{id}")
    public ModelMap delete(@PathVariable("id") Integer id);
    @GetMapping("/permission/willEditPermission/{id}")
    public ModelMap willEditPermission(@PathVariable("id") Integer id) ;
    @GetMapping("/permission/find/{id}")
    public ModelMap find(@PathVariable("id") Integer id);
    @GetMapping("/permission/findAll")
    public ModelMap findAll();
    @RequestMapping("/permission/manage")
    public ModelMap manage(@RequestParam("pageNo") final Integer pageNo, @RequestParam("pageSize") final Integer pageSize) ;
}
```

熔断处理的代码如下：

```java
package com.rc.user.clientapi;
//导包语句省略
...
@Slf4j
@Component
public class RcUserClientApiFallback implements FallbackFactory<RcUserClientApi> {
    @Override
    public RcUserClientApi create(Throwable throwable) {
        return new RcUserClientApi(){
            @Override
            public ModelMap register(RcUserQo rcUser) {
                return defaultFallbackResult();
            }
            @Override
            public ModelMap add(RcUserQo rcUser) {
                return defaultFallbackResult();
            }
            @Override
            public ModelMap findByUserName(String userName) {
                return defaultFallbackResult();
            }
            @Override
            public ModelMap personalEdit(RcUserQo rcUser) {
                return defaultFallbackResult();
            }
```

```java
            @Override
            public ModelMap edit(RcUserQo rcUser) {
                return defaultFallbackResult();
            }
            @Override
            public ModelMap forFindPwd(String userName, String userEmail, String newPwd) {
                return defaultFallbackResult();
            }
            @Override
            public ModelMap willEditPwd() {
                return defaultFallbackResult();
            }
            @Override
            public ModelMap editPwd(Integer userId, String userPwd, String newPwd) {
                return defaultFallbackResult();
            }
            @Override
            public ModelMap delete(Integer id) {
                return defaultFallbackResult();
            }
            @Override
            public ModelMap deleteBatch(String ids) {
                return defaultFallbackResult();
            }
            @Override
            public ModelMap willEditRole(Integer id) {
                return defaultFallbackResult();
            }
            @Override
            public ModelMap editRole(Integer id, Integer[] roleIds) {
                return defaultFallbackResult();
            }
            @Override
            public ModelMap find(Integer id) {
                return defaultFallbackResult();
            }
            @Override
            public ModelMap manage(String userName, Integer pageNo, Integer pageSize) {
                return defaultFallbackResult();
            }
            protected ModelMap defaultFallbackResult() {
                log.debug("请求服务失败："+throwable.getMessage());
                ModelMap map=new ModelMap();
                map.put("code","0");
                map.put("date","请求服务失败");
                return map;
            }
        };
    }
}
```

6.5 视图微服务设计

为了便于部署和灵活调整应用，视图被存放在独立的微服务中，而与业务直接相关的微服务，即 Web UI 微服务被存放在业务相关项目中心中。例如，user-web 微服务被存放在 user-microservice 项目中；统一入口的视图需要单独建立一个项目并命名为 client-web。

6.5.1 Thymeleaf 技术

1. Thymeleaf 简介

Thymeleaf 是一个用于 Web 和独立 Java 环境的模板引擎，能够处理 HTML、XML、JavaScript、CSS 甚至纯文本。它能轻易地与 Spring MVC 等 Web 框架进行集成并作为 Web 应用的模板引擎。与其他模板引擎（如 FreeMarker）相比，Thymeleaf 最大的特点是能够直接在浏览器中打开并正确显示模板页面，而不需要启动整个 Web 应用，因此更方便进行前端设计。Thymeleaf 是 Spring Boot 官方推荐使用的模板。

Thymeleaf 3.0 是一个彻底重构的模板引擎，极大地减少了内存占用，提升了性能和并发性，避免了 2.1 版本因大量的输出标记的集合而产生的资源占用。

Thymeleaf 具有如下特征。

- 在有网络和无网络的环境下皆可运行，即它可以使美工在浏览器中查看静态页面效果，也可以使程序员在服务器中查看带数据的动态页面效果。
- 开箱即用。它提供了标准和 Spring 标准两种方言，可以直接套用模板实现 JSTL、ONGL 表达式效果，避免每天套模板、改 JSTL 和改标签的困扰。同时，开发人员可以扩展和创建自定义的方言。
- 提供了 Spring 标准方言和一个与 Spring MVC 完美集成的可选模块，可以快速地实现表单绑定、属性编辑器和国际化等功能。

使用 Thymeleaf 需要引入如下依赖：

```
<dependency>
    <groupId>org.springframework.boot</groupId>
    <artifactId>spring-boot-starter-thymeleaf</artifactId>
</dependency>
```

同时，在配置文件中增加如下配置：

```
#是否开启缓存，开发时可以设置为 false，默认为 true
spring.thymeleaf.cache=true
#模板文件编码
spring.thymeleaf.encoding=UTF-8
#模板文件位置
spring.thymeleaf.prefix=classpath:/templates/
#Content-Type 配置
spring.thymeleaf.servlet.content-type=text/html
#模板文件后缀
```

spring.thymeleaf.suffix=.html

2．基础语法

1）Thymeleaf 页面的基本结构

在使用 Thymeleaf 时，需要将文件放入 classpath:/templates/目录中，文件后缀是.html。在 HTML 页面中引入 Thymeleaf 命名空间，此时在 HTML 模板文件中，动态的属性使用 th:命名空间修饰。

2）变量表达式${...}

变量表达式通过${...}进行取值，这与 ONGL 表达式语法一致。例如：

```
<p th:text="'Hello! , ' + ${name} + '!'">3333</p>
<span th:text="${book.author.name}">
```

3）选择变量表达式*{...}

选择变量表达式用于访问对象的属性，并通过 th:object 绑定对象。例如：

```
<div th:object="${session.user}">
    <p>Name: <span th:text="*{firstName}">Sebastian</span>.</p>
    <p>Surname: <span th:text="*{lastName}">Pepper</span>.</p>
    <p>Nationality: <span th:text={nationality}">Saturn</span>.</p>
</div>
```

等价于：

```
<div>
    <p>Name: <span th:text="${session.user.firstName}">Sebastian</span>.</p>
    <p>Surname: <span th:text="${session.user.lastName}">Pepper</span>.</p>
    <p>Nationality: <span th:text="${session.user.nationality}">Saturn</span>.</p>
</div>
```

4）消息表达式#{...}

消息表达式用于展示静态资源的内容，如 i18n 属性配置文件。例如，新建/WEB-INF/templates/home.properties 文件，内容如下：

```
home.welcome=this messages is from home.properties!
```

使用消息表达式：

```
<p th: text=" #{home.welcome}" >This text will not be show! </p>
```

5）URL 表达式@{...}

URL 表达式用于配合 link、src、href 使用，类似的标签有 th:href 和 th:src。例如：

带参数的 URL：

```
<a th:href="@{/order/details(id=${orderId},type=${orderType})}">...</a>
```

相对地址：

```
<a th:href="@{/documents/report}">...</a>
```

绝对地址：

```
<a th:href="@{http://www.mycompany.com/main}">...</a>
```

6）文本替换

字符串连接：

```
<span th:text="'Welcome to our application, ' + ${user.name} + '!'">
```

字面量替换（只能包含表达式变量，不能包含条件判断等）：

```
<span th:text="|Welcome to our application, ${user.name}!|">
```

7）条件判断

使用 th:if 和 th:unless 属性进行条件判断。th:unless 与 th:if 恰好相反，只有表达式中的条件不成立，才会显示其内容。例如：

```
<a th:href="@{/login}" th:unless=${session.user != null}>Login</a>
```

对于多选择情况，也可以使用 th:switch。例如：

```
<div th:switch="${user.role}">
    <p th:case="'admin'">User is an administrator</p>
    <p th:case="#{roles.manager}">User is a manager</p>
    <p th:case="*">User is some other thing</p>
</div>
```

8）数据遍历

使用 th:each 进行数据遍历，例如：

```
<tr th:each="emp : ${empList}">
    <td th:text="${emp.id}">1</td>
    <td th:text="${emp.name}">海</td>
    <td th:text="${emp.age}">18</td>
</tr>
```

6.5.2　Web UI 微服务设计

Web UI 微服务通过使用 ClientApi 接口与 REST API 微服务进行数据通信，并以合适的视图将数据呈现给用户或允许用户交互。这里以 user-web 为例进行说明。

1．创建 user-web 模块

在用户微服务项目 user-microservice 中创建 user-web 模块，并引入如下依赖：

```xml
<dependency>
    <groupId>org.springframework.boot</groupId>
    <artifactId>spring-boot-starter</artifactId>
</dependency>
<dependency>
    <groupId>org.springframework.boot</groupId>
    <artifactId>spring-boot-starter-web</artifactId>
</dependency>
<dependency>
    <groupId>org.springframework.cloud</groupId>
    <artifactId>spring-cloud-starter-openfeign</artifactId>
</dependency>
<dependency>
    <groupId>org.springframework.cloud</groupId>
    <artifactId>spring-cloud-starter-netflix-eureka-client</artifactId>
</dependency>
<dependency>
    <groupId>org.springframework.cloud</groupId>
    <artifactId>spring-cloud-starter-config</artifactId>
</dependency>
<dependency>
```

```xml
        <groupId>org.springframework.boot</groupId>
        <artifactId>spring-boot-starter-thymeleaf</artifactId>
</dependency>
<dependency>
        <groupId>com.rc</groupId>
        <artifactId>user-clientapi</artifactId>
        <version>0.1-SNAPSHOT</version>
</dependency>
<dependency>
        <groupId>com.rc</groupId>
        <artifactId>company-clientapi</artifactId>
        <version>0.1-SNAPSHOT</version>
</dependency>
```

在上述依赖中引用 user-clientapi 的目的是通过 FeignClient 接口调用 REST API。由于在添加用户时需要企业列表，需要调用企业 REST API，因此引用了 company-clientapi 依赖。

2. 设计 Web Controller

这里的 Controller 用于接收视图端请求，并通过 FeignClient 接口调用 REST API。以用户 Controller 为例，代码如下：

```java
package com.rc.user.controller;
//导包语句省略
...
@Controller
@RequestMapping("/user")
public class RcUserController extends RcBaseController {
    @Resource
    private RcUserClientApi rcUserClientApi;
    @Resource
    private RcCompanyClientApi rcCompanyClientApi;
    @GetMapping("/willAdd")
    public ModelAndView willAdd() {
        ModelMap map= rcCompanyClientApi.list();
        if (map.get("code").toString().equals("1")) {
            return new ModelAndView("/user/add", "companylist", map.get("data"));
        } else {
            return new ModelAndView("/error", "msg", map.get("data"));
        }
    }
    @GetMapping("/willRegister")
    public String willRegister() {
        return "/user/register";
    }
    @PostMapping("/register")
    public ModelAndView register(RcUserQo user) {
        ModelMap map = rcUserClientApi.register(user);
        if (map.get("code").toString().equals("1")) {
            return new ModelAndView("redirect:http://localhost:9099");
```

```java
        } else {
            return new ModelAndView("/error", "msg", map.get("data"));
        }
    }
    @PostMapping("/add")
    public ModelAndView add(RcUserQo user) {
        ModelMap map = rcUserClientApi.add(user);
        if (map.get("code").toString().equals("1")) {
            return new ModelAndView("redirect:/user/manage");
        } else {
            return new ModelAndView("/error", "msg", map.get("data"));
        }
    }

    @GetMapping("/personalWillEdit")
    public ModelAndView personalWillEdit() {
        Integer userId=5;
        ModelMap map = rcUserClientApi.find(userId);
        if (map.get("code").toString().equals("1")) {
            return new ModelAndView("/user/personalEdit", "rcUser", map.get("data"));
        } else {
            return new ModelAndView("/error", "msg", map.get("data"));
        }
    }
    @PostMapping("/personalEdit")
    public ModelAndView personalEdit(RcUserQo rcUser) {
        ModelMap map = rcUserClientApi.personalEdit(rcUser);
        if (map.get("code").toString().equals("1")) {
            return new ModelAndView("redirect:http://localhost:9099/center");
        } else {
            return new ModelAndView("/error", "msg", map.get("data"));
        }
    }
    @GetMapping("/willEdit/{id}")
    public ModelAndView willEdit(@PathVariable("id") Integer id) {
        ModelMap map1 = rcUserClientApi.find(id);
        ModelMap map2= rcCompanyClientApi.list();
        if (map1.get("code").toString().equals("1")&&map2.get("code").toString().equals("1")) {
            ModelMap map=new ModelMap();
            map.put("rcUser",map1.get("data"));
            map.put("companylist",map2.get("data"));
            return new ModelAndView("/user/edit","map",map);
        } else {
            return new ModelAndView("/error", "msg", map1.get("data"));
        }
    }
    @PostMapping("/edit")
    public ModelAndView edit(RcUserQo user) {
        ModelMap map = rcUserClientApi.edit(user);
```

```java
            if (map.get("code").toString().equals("1")) {
                return new ModelAndView("redirect:/user/manage");
            } else {
                return new ModelAndView("/error", "msg", map.get("data"));
            }
        }
        @PostMapping("/forFindPwd")
        public ModelAndView forFindPwd(String userName, String userEmail, String newPwd) {
            ModelMap map = rcUserClientApi.forFindPwd(userName, userEmail, newPwd);
            if (map.get("code").toString().equals("1")) {
                return new ModelAndView("redirect:http://localhost:5777/login");
            } else {
                return new ModelAndView("/error", "msg", map.get("data"));
            }
        }
        @GetMapping("/willEditPwd")
        public ModelAndView willEditPwd() {
            Integer userId = 5;
            return new ModelAndView("/user/editPwd", "rcUser", userId);
        }
        @PostMapping("/editPwd")
        public ModelAndView editPwd(Integer userId, String userPwd, String newPwd) {
            ModelMap map = rcUserClientApi.editPwd(userId, userPwd, newPwd);
            if (map.get("code").toString().equals("1")) {
                return new ModelAndView("redirect:http://localhost:8084/web/center");
            } else {
                return new ModelAndView("/error", "msg", map.get("data"));
            }
        }
        @GetMapping("/delete/{id}")
        public ModelAndView delete(@PathVariable("id") Integer id) {
            ModelMap map = rcUserClientApi.delete(id);
            if (map.get("code").toString().equals("1")) {
                return new ModelAndView("redirect:/center/user/manage");
            } else {
                return new ModelAndView("/error", "msg", map.get("data"));
            }
        }
        @PostMapping("/delete/batch")
        public ModelAndView deleteBatch(String ids) {
            ModelMap map = rcUserClientApi.deleteBatch(ids);
            if (map.get("code").toString().equals("1")) {
                return new ModelAndView("redirect:/user/manage");
            } else {
                return new ModelAndView("/error", "msg", map.get("data"));
            }
        }
        @GetMapping("/willEditRole/{id}")
        public ModelAndView willEditRole(@PathVariable("id") Integer id) {
```

```java
            ModelMap map = rcUserClientApi.willEditRole(id);
            if (map.get("code").toString().equals("1")) {
                return new ModelAndView("/user/editRole", "map", map.get("data"));
            } else {
                return new ModelAndView("/error", "msg", map.get("data"));
            }
        }
        @PostMapping("/editRole/{id}")
        public ModelAndView editRole(@PathVariable("id") Integer id, Integer[] roleIds) {
            ModelMap map = rcUserClientApi.editRole(id, roleIds);
            if (map.get("code").toString().equals("1")) {
                return new ModelAndView("redirect:/user/manage");
            } else {
                return new ModelAndView("/error", "msg", map.get("data"));
            }
        }
        @GetMapping("/show/{id}")
        public ModelAndView show(@PathVariable("id") Integer id) {
            ModelMap map = rcUserClientApi.find(id);
            if (map.get("code").toString().equals("1")) {
                return new ModelAndView("/user/show");
            } else {
                return new ModelAndView("/error", "msg", map.get("data"));
            }
        }
        @RequestMapping("/manage")
        public ModelAndView manage(String userName, Integer pageNo, Integer pageSize) {
            if (pageNo == null) pageNo = 1;
            if (pageSize == null) pageSize = 10;
            ModelMap map = rcUserClientApi.manage(userName, pageNo, pageSize);
            if ((map.get("code")).toString().equals("1")) {
                LinkedHashMap map1=(LinkedHashMap)map.get("data");
                PageHelper pageHelper=new PageHelper(Long.valueOf(map1.get("total").toString()),
            pageNo,pageSize);
                map1.put("pageHelper",pageHelper);
                return new ModelAndView("/user/manage", "map", map1);
            } else {
                return new ModelAndView("/error", "msg", map.get("data"));
            }
        }
}
```

3. 视图设计

微服务模块中的视图可以是 PC Web，也可以是移动 Web，本节仅介绍 PC Web。PC Web 的设计方法与单体结构中的设计方法基本相同，所不同的是，在每个业务工程项目中，Web 微服务模块所涉及的视图仅与这个业务相关。例如，user-web 所涉及的视图仅包括用户、角色和权限 3 方面的视图。这里仍然需要采用 Thymeleaf 技术。因篇幅限制，这里仅给出部分用户视图的设计。

1）用户管理视图设计

```html
<!DOCTYPE html>
<html xmlns:th="http://www.thymeleaf.org">
<head>
    <meta http-equiv="Content-Type" content="text/html; charset=UTF-8"/>
    <title>用户管理</title>
    <script type="text/javascript" th:src="@{/js/jquery.min.js}"></script>
    <script type="text/javascript" th:src="@{/js/jquery-ui.min.js}"></script>
    <link rel="stylesheet" th:href="@{/css/jquery-ui.min.css}" type="text/css"/>
    <link rel="stylesheet" th:href="@{/css/jquery-ui.theme.min.css}" type="text/css"/>
    <link rel="stylesheet" th:href="@{/css/all.css}" type="text/css"/>
    <script>
        function showDialog(url) {
            $("<div></div>").load(url).dialog({
                modal: true,
                width: 700,
                height: 400,
                zIndex: 1000,
                buttons: {
                    "确定": function () {
                        $(this).dialog("close");
                    }
                }
            });
        }
    </script>
</head>
<body>
<div class="navigator">
    <a href="http://localhost:8084/web/sysmanage" target="_top">系统管理</a> > 用户管理
</div>
<form method="post" th:action="@{/user/manage}">
    <a class="button add" th:href="@{/user/willAdd}">添加</a>
    <a class="button delete" href="JavaScript:document.getElementById('form2').submit()">删除</a>
    <input type="hidden" name="pageNo" th:value="${map['pageNo']}"/>
    <input type="hidden" name="pageSize" th:value="${map['pageSize']}"/>
    用户名：<input type="text" name="userName" size="20" th:value="${map['userName']}"/>
    <input class="button find" type="button" value="查询" onclick="goPage(1);"/>
</form>
<form id="form2" method="post" th:action="@{/user/delete/batch}">
    <table class="mytable1" align="center" cellpadding="4" cellspacing="0">
        <tr>
            <th><input id="sel" type="checkbox" value="${u.userId}"/></th>
            <th>用户类型</th>
            <th>用户名</th>
            <th>真实名</th>
            <th>角色</th>
            <th>所属企业</th>
            <th width="100" class="right">操作</th>
```

```html
            </tr>
            <tr th:each="u:${map['rows']}">
                <td><input type="checkbox" name="ids" th:value="${u.userId}"/></td>
                <td th:text="${u.userType}"></td>
                <td><a th:href="@{http://localhost:8084/person-web/person/showByUserId/{id}(id=${u.userId})}" th:text="${u.userRealname}" onclick="showDialog(this.href);return false"></a></td>
                <td th:text="${u.userRealname}"></td>
                <td>
                    <span th:each="r:${u.roles}" th:text="${r.roleName+' '}"></span>
                </td>
                <td th:text="${u.comName}"></td>
                <td class="right">
                    <a th:href="@{/user/willEdit/{id}(id=${u.userId})}">修改</a> 
                    <a th:href="@{/user/willEditRole/{id}(id=${u.userId})}">设置</a> 
                    <a th:href="@{/user/delete/{id}(id=${u.userId})}">删除</a>
                </td>
            </tr>
        </table>
        <div>[(${map['pageHelper'].pageBar})]</div>
    </form>
</body>
</html>
```

2）添加用户视图设计

```html
<!DOCTYPE html>
<html xmlns:th="http://www.thymeleaf.org">
<head>
    <meta http-equiv="Content-Type" content="text/html; charset=UTF-8">
    <title>添加用户</title>
    <link rel="stylesheet" th:href="@{/css/all.css}" type="text/css"/>
</head>
<body>
<div class="navigator"><a href="http://localhost:8084/web/sysmanage" target="_top">系统管理</a> > <a th:href="@{/user/manage}">用户管理</a>  >添加用户</div>
<form method="POST" th:action="@{/user/add}">
    <table class="mytable2" align="center" cellspacing="0" cellpadding="2">
        <tr>
            <th>用户名</th>
            <td><input type="text" name="userName" size="18"></td>
        </tr>
        <tr>
            <th>用户类型</th>
            <td>
                <input type="radio" value="系统用户" name="userType" checked
                onclick="document.getElementById('comSelect').style.display='none'"/>系统用户
                 <input type="radio" value="企业用户" name="userType"
                onclick="document.getElementById('comSelect').style.display=''"/>企业用户
            </td>
        </tr>
        <tr id="comSelect" style="display: none">
```

```html
            <th>所属企业</th>
            <td>
                <select size="1" name="comId" onchange="document.getElementsByName('comName')[0].value=this.options[this.selectedIndex].text">
                    <option value="" selected>请选择</option>
                    <option th:each="c:${companylist}" th:value="${c.comId}" th:text="${c.comName}"></option>
                </select>
                <input type="text" name="comName" size="20"/>
            </td>
        </tr>
        <tr>
            <th>密码</th>
            <td><input type="password" name="userPwd" size="18"></td>
        </tr>
        <tr>
            <th>确认密码</th>
            <td><input type="password" name="userPwd1" size="18"></td>
        </tr>
        <tr>
            <th>真实名</th>
            <td><input type="text" name="userRealname" size="18"></td>
        </tr>
        <tr>
            <th>邮箱</th>
            <td><input type="text" name="userEmail" size="60"></td>
        </tr>
    </table>
    <p align="center">
        <input class="button ok" type="submit" value=" 提 交 ">
        <input class="button cancel" type="button" value=" 取 消 " onclick="window.history.back();">
    </p>
</form>
</body>
</html>
```

3）修改用户视图设计

```html
<!DOCTYPE html>
<html xmlns:th="http://www.thymeleaf.org">
<head>
    <meta http-equiv="Content-Type" content="text/html; charset=UTF-8">
    <title>修改用户</title>
    <script type="text/javascript" th:src="@{/js/jquery.min.js}"></script>
    <link rel="stylesheet" th:href="@{/css/all.css}" type="text/css"/>
    <script type="text/javascript" th:inline="javascript">
        userType = [[${map['rcUser'].userType}]];
        comId = [[${map['rcUser'].comId}]];
        $(function () {
            if(userType=="企业用户"){
                $("select[name='comId']").val(comId);
```

```html
                    $("#comSelect").show();
                }
                $("input:radio[value='" + userType + "']").prop("checked", true);
            });
        </script>
    </head>
    <body>
        <div class="navigator"><a href="http://localhost:8084/web/sysmanage" target="_top">系统管理</a> > <a th:href="@{/user/manage}">用户管理</a> > 修改用户</div>
        <form method="POST" th:action="@{/user/edit}">
            <input type="hidden" name="userId" th:value="${map['rcUser'].userId}"/>
            <table class="mytable2" align="center" cellspacing="0" cellpadding="2">
                <tr>
                    <th>用户名</th>
                    <td>
                        <input type="text" name="userName" size="18"
                            th:value="${map['rcUser'].userName}"/></td>
                </tr>
                <tr>
                    <th>用户类型</th>
                    <td>
                        <input type="radio" value="系统用户" name="userType"
                        onclick="document.getElementById('comSelect').style.display='none'"/>系统用户
                         <input type="radio" value="企业用户" name="userType"
                        onclick="document.getElementById('comSelect').style.display=''"/>企业用户
                         <input type="radio" value="个人用户" name="userType"
                        onclick="document.getElementById('comSelect').style.display='none'"/>个人用户
                    </td>
                </tr>
                <tr id="comSelect" style="display: none">
                    <th>所属企业</th>
                    <td>
                        <select size="1" name="comId" onchange="document.getElementsByName('comName')[0].value=this.options[this.selectedIndex].text">
                            <option value="" selected>请选择</option>
                            <option th:each="c:${map['companylist']}" th:value="${c.comId}" th:text="${c.comName}"></option>
                        </select>
                        <input type="text" name="comName" size="20"/>
                    </td>
                </tr>
                <tr>
                    <th>真实名</th>
                    <td><input type="text" name="userRealname" size="18"
                            th:value="${map['rcUser'].userRealname}"/></td>
                </tr>
                <tr>
                    <th>邮箱</th>
                    <td><input type="text" name="userEmail" size="60"
```

```html
                          th:value="${map['rcUser'].userEmail}"/></td>
            </tr>
        </table>
        <p align="center">
            <input class="button ok" type="submit" value=" 提 交 "/>
            <input class="button cancel" type="button" value=" 取 消 " onclick="self.history.back();"/>

        </p>
    </form>
</body>
</html>
```

6.5.3 统一入口微服务设计

由于不同的业务功能被分散在不同的微服务中,因此用户需要统一的入口,从而方便使用。下面创建独立的项目 web-microservice,使其包含统一入口界面。

1. 创建工程

创建 web-microservice 项目,并引入如下依赖:

```xml
<dependency>
    <groupId>org.springframework.boot</groupId>
    <artifactId>spring-boot-starter-web</artifactId>
</dependency>
<dependency>
    <groupId>org.springframework.boot</groupId>
    <artifactId>spring-boot-devtools</artifactId>
    <scope>runtime</scope>
    <optional>true</optional>
</dependency>
<dependency>
    <groupId>org.springframework.cloud</groupId>
    <artifactId>spring-cloud-starter-netflix-eureka-client</artifactId>
</dependency>
<dependency>
    <groupId>org.springframework.boot</groupId>
    <artifactId>spring-boot-configuration-processor</artifactId>
    <optional>true</optional>
</dependency>
<dependency>
    <groupId>org.springframework.cloud</groupId>
    <artifactId>spring-cloud-starter-openfeign</artifactId>
</dependency>
<dependency>
    <groupId>org.springframework.boot</groupId>
    <artifactId>spring-boot-starter-thymeleaf</artifactId>
</dependency>
<dependency>
    <groupId>com.rc</groupId>
```

```xml
        <artifactId>person-clientapi</artifactId>
        <version>0.0.1-SNAPSHOT</version>
</dependency>
<dependency>
        <groupId>com.rc</groupId>
        <artifactId>company-clientapi</artifactId>
        <version>0.0.1-SNAPSHOT</version>
</dependency>
<dependency>
        <groupId>com.rc</groupId>
        <artifactId>job-clientapi</artifactId>
        <version>0.0.1-SNAPSHOT</version>
</dependency>
<dependency>
        <groupId>org.springframework.boot</groupId>
        <artifactId>spring-boot-starter-test</artifactId>
        <scope>test</scope>
</dependency>
```

2．配置文件 bootstrap.properties

```
spring.application.name=web-client
eureka.client.serviceUrl.defaultZone=http://localhost:8099/eureka/
server.port=9099
server.servlet.context-path=/web
ribbon.eureka.enabled=true
feign.client.config.default.connect-timeout=60000
feign.client.config.default.read-timeout=60000
spring.main.allow-bean-definition-overriding=true
spring.thymeleaf.cache=false
spring.thymeleaf.encoding=UTF-8
spring.thymeleaf.mode=HTML
spring.thymeleaf.prefix=classpath:/templates
spring.thymeleaf.suffix=.html
```

3．创建一个启动程序

```java
package com.rc.webclient;
...//导包语句省略
@SpringBootApplication(exclude = DataSourceAutoConfiguration.class)
@EnableDiscoveryClient
@EnableFeignClients(basePackages= {"com.rc.person.clientapi","com.rc.job.clientapi","com.rc.company.clientapi"})
public class WebClientApplication {
    public static void main(String[] args) {
        SpringApplication.run(WebClientApplication.class, args);
    }
}
```

4．创建一个控制类

```java
package com.rc.webclient.controller;
...//导包语句省略
```

```java
@Controller
@RequestMapping("/")
public class WebController {
    @Resource
    private RcPersonClientApi rcPersonClientApi;
    @Resource
    private RcJobClientApi rcJobClientApi;
    @Resource
    private RcCompanyClientApi rcCompanyClientApi;
    @RequestMapping("/center")
    public String center(){
        return "/center";
    }
    @RequestMapping("/sysmanage")
    public String sysmanage(){
        return "/sysmanage";
    }
    @RequestMapping("/companymanage")
    public String companymanage(){
        return "/companymanage";
    }
    @RequestMapping("/first1")
    public String first1(){
        return "/first1";
    }
    @RequestMapping("/")
    public String home(){
        return "/index";
    }
    @RequestMapping("/first")
    public ModelAndView first() throws Exception {
        ModelMap map = new ModelMap();
        ModelMap map1 = rcJobClientApi.findByNum(5);
        ModelMap map2 = rcPersonClientApi.findByNum(5);
        ModelMap map3 = rcCompanyClientApi.findByNum(5);
        map.put("jobs", map1.get("data"));
        map.put("persons", map2.get("data"));
        map.put("companies", map3.get("data"));
        return new ModelAndView("/first", "map", map);
    }
}
```

5. 建立视图

1）系统首页

在微服务架构下调整系统首页，将登录部分单独拿出去，如图 6-10 所示。

第 6 章 微服务架构与 Spring Cloud

图 6-10 系统首页

示例代码如下：

```
<!DOCTYPE html>
<html xmlns:th="http://www.thymeleaf.org">
<head>
    <meta http-equiv="Content-Type" content="text/html; charset=UTF-8"/>
    <title>网上人才中心系统</title>
    <link rel="stylesheet" th:href="@{/css/all.css}" type="text/css">
</head>
<body>
<div id="container">
    <div id="header"></div>
    <div id="menu">
        <ul>
            <li><a th:href="@{/}" target="_top">【系统首页】</a></li>
            <li><a href="http://localhost:8084/news-web/news/browse" target="content">【站内新闻】
            </a></li>
            <li><a href="http://localhost:8084/job-web/job/browse" target="content">【招聘信息】
            </a></li>
            <li><a href="http://localhost:8084/company-web/company/browse" target="content">【企业信息】</a></li>
            <li><a href="http://localhost:8084/person-web/person/browse" target="content">【人才信
```

```
            息】</a></li>
                <li><a th:href="@{/center}" target="_top">【个人中心】</a></li>
            </ul>
        </div>
        <div id="pagebody">
            <iframe id="content" name="content" frameborder="0" width="100%" height="100%"
        th:src="@{/first}">
            </iframe>
        </div>
        <div id="footer">
            <hr/>
            Copyright &copy; 2020-2030 YSL.All Rights Reserved.
        </div>
    </div>
    </body>
    </html>
```

2）个人中心

【个人中心】界面如图 6-11 所示。

图 6-11 【个人中心】界面

示例代码如下：

```html
<!DOCTYPE html>
<html xmlns:th="http://www.thymeleaf.org">
<head>
    <meta http-equiv="Content-Type" content="text/html; charset=UTF-8"/>
    <title>网上人才中心系统</title>
    <link rel="stylesheet" href="./css/all.css" type="text/css"/>
</head>
<body>
<div id="container">
    <div id="header"></div>
    <div id="pagebody">
        <div id="left">
            <div class="title">个人中心</div>
            <div class="leftmenu">
                <a href="http://localhost:8084/user-web/user/personalWillEdit" target="content">个人修改</a>
                <a href="http://localhost:8084/person-web/person/willSetInfo" target="content">完善信息</a>
                <a href="http://localhost:8084/job-web/application/browse" target="content">我的申请</a>
                <a href="http://localhost:8084/user-web/user/willEditPwd" target="content">修改密码</a>
                <a href="http://localhost:8084/web" target="_top">返回首页</a>
            </div>
        </div>
        <div id="right">
            <iframe id="content" name="content" frameborder="0" style="width:100%;height:300px" th:src="@{/first1}"></iframe>
        </div>
    </div>
    <div id="footer">
        <hr>
        Copyright &copy; 2020-2030 YSL.All Rights Reserved.
    </div>
</div>
</body>
</html>
```

3）系统管理

【系统管理】界面如图6-12所示。

Java EE 企业级应用开发技术研究

图 6-12 【系统管理】界面

示例代码如下：

```html
<!DOCTYPE html>
<html xmlns:th="http://www.thymeleaf.org">
<head>
    <meta http-equiv="Content-Type" content="text/html; charset=UTF-8"/>
    <title>网上人才中心系统</title>
    <link rel="stylesheet" th:href="@{/css/all.css}" type="text/css"/>
</head>
<body>
<div id="container">
    <div id="header"></div>
    <div id="pagebody">
        <div id="left">
            <div class="title">系统管理</div>
            <div class="leftmenu">
                <a href="http://localhost:8084/company-web/company/manage" target="content">用户管理</a>
                <a href="http://localhost:8084/user-web/user/manage" target="content">用户管理</a>
                <a href="http://localhost:8084/user-web/permission/manage" target="content">权限管理</a>
                <a href="http://localhost:8084/user-web/role/manage" target="content">角色管理</a>
                <a href="http://localhost:8084/logout">退出系统</a>
```

```
            </div>
        </div>
        <div id="right">
            <iframe id="content" name="content" frameborder="0" style="width:100%;height:300px"
            th:src="@{/first1}"></iframe>
        </div>
    </div>
    <div id="footer">
        <hr/>
        Copyright &copy; 2020-2030 YSL.All Rights Reserved.
    </div>
</div>
</body>
```

4）企业管理

【企业管理】界面如图 6-13 所示。

图 6-13 【企业管理】界面

示例代码如下：

```
</html>
<!DOCTYPE html>
<html xmlns:th="http://www.thymeleaf.org">
<head>
```

```html
        <meta http-equiv="Content-Type" content="text/html; charset=UTF-8"/>
        <title>网上人才中心系统</title>
        <link rel="stylesheet" th:href="@{/css/all.css}" type="text/css"/>
</head>
<body>
<div id="container">
    <div id="header"></div>
    <div id="pagebody">
        <div id="left">
            <div class="title">企业管理</div>
            <div class="leftmenu">
                <a href="http://localhost:8084/job-web/job/manage" target="content">招聘管理</a>
                <a href="http://localhost:8084/job-web/application/manage" target="content">应聘管理</a>
                <a href="http://localhost:8084/logout}">退出系统</a>
            </div>
        </div>
        <div id="right">
            <iframe id="content" name="content" frameborder="0" style="width:100%;height:300px" th:src="@{/first1}"></iframe>
        </div>
    </div>
    <div id="footer">
        <hr/>Copyright &copy; 2020-2030 YSL.All Rights Reserved.
    </div>
</div>
</body>
</html>
```

第 7 章　在微服务架构中整合 OAuth2

本章主要介绍 OAuth2 基本原理、JWT 技术、在微服务架构中实现 SSO，并结合网上人才中心系统介绍授权服务器模块设计及实现微服务应用访问控制的方法。

7.1　基于 OAuth2 实现 SSO 的原理

7.1.1　OAuth2 基本原理

OAuth（Open Authorization，开放授权）为用户资源的授权定义了一个安全、开放及简单的标准，其核心原理是通过各种认证手段认证用户身份，并颁发一个 token（令牌），使得第三方应用可以使用该令牌在限定时间、限定范围内访问指定资源。每一个令牌会授权一个特定的应用在特定的时段内访问特定的资源。这样，OAuth 就可以让用户通过授权第三方应用来灵活地访问存储在另外一些资源服务器的特定信息，而非所有内容。

目前，主流的 QQ、微信等第三方应用的授权登录方式都是基于 OAuth2 实现的。OAuth 2.0 是 OAuth 协议的延续版本，但不向后兼容 OAuth 1.0，即完全废止了 OAuth 1.0。

1．OAuth2 角色

OAuth 2 标准中定义了以下几种角色。

资源所有者（Resource Owner）：代表授权客户端访问本身资源信息的用户。客户端访问用户账户的权限仅限于用户授权的"范围"。

客户端（Client）：代表意图访问受限资源的第三方应用。在访问实现之前，客户端必须先经过用户授权，并且获得的授权凭证将进一步由授权服务器进行验证。

授权服务器（Authorization Server）：授权服务器用来验证用户提供的信息是否正确，并返回一个令牌给第三方应用。

资源服务器（Resource Server）：资源服务器是向用户提供资源的服务器，如头像、照片、视频等。

2．OAuth2 授权流程

一个大致的 OAuth2 授权流程如图 7-1 所示。

步骤 1：客户端（第三方应用）向用户请求授权。

步骤 2：用户单击客户端所呈现的服务授权页面上的【同意授权】按钮后，会返回一个授权许可凭证给客户端。

步骤 3：客户端凭借授权许可凭证向授权服务器申请令牌。

步骤 4：授权服务器在验证信息无误后，发放令牌给客户端。

步骤 5：客户端凭借令牌向资源服务器申请资源。
步骤 6：资源服务器在验证令牌无误后开放资源。

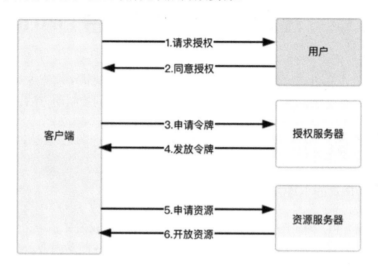

图 7-1 OAuth2 授权流程

3. OAuth2 授权模式

OAuth2 授权模式共分为 4 种。

授权码模式：授权码模式（Authorization Code）是功能最完整、流程最严谨的授权模式。它的特点是通过客户端与授权服务器进行交互。国内常见的第三方应用登录功能基本都使用这种模式。

简化模式：简化模式不需要客户端参与，可以直接在浏览器中向授权服务器申请令牌。在一般情况下，如果网站是纯静态页面，则可以采用这种模式。

密码模式：密码模式是用户将用户名和密码直接告诉客户端，让客户端使用这些信息向授权服务器申请令牌。这需要用户对客户端高度信任，如客户端和服务的提供商是同一家公司。

客户端模式：客户端模式是指客户端以自己的名义而不是以用户的名义向资源所有者请求授权。严格来说，客户端模式并不能算作 OAuth2 针对要解决的问题的一种解决方案，但是，对于开发者而言，在一些前后端分离的应用或为移动端提供的授权服务器上使用这种模式还是非常方便的。

7.1.2 JWT 概述

OAuth2 使用 token 验证用户登录的合法性，但是 token 最大的问题是不携带用户信息，使得资源服务器无法在本地进行验证。在每次对资源进行访问时，资源服务器都需要向授权服务器发起请求，一方面是为了验证 token 的有效性；另一方面是为了获取 token 对应的用户信息。如果有大量的此类请求，则处理效率会很低，而且授权服务器会变成一个中心节点，这在分布式架构下会严重影响性能。

JWT（JSON Web Token）就是在这样的背景下诞生的。JWT 是一种开放的标准（RFC 7519），旨在将各个主体的信息包装为 JSON 对象。主体信息是通过数字签名进行加密和验证的。通常使用 HMAC 算法或 RSA（公钥/私钥的非对称性加密）算法对 JWT 进行签名。

普通的 OAuth2 颁发的是一串随机 Hash 字符串，本身并无意义。而 JWT 使用了一种特殊格式的 token。这种 token 是有特定含义的，可以分为 3 部分：Header（头部）、Payload（载荷）、Signature（签名）。授权服务器通过对称或非对称的加密方式使用 Payload 生成 Signature，并在 Header 中声明签名方式。这 3 部分均使用 Base64 进行编码，并使用"."进行分隔。一个典型的 JWT 格式的 token 类似于 xxxxx.yyyyy.zzzzz。

通过这种方式，JWT 可以实现分布式的 token 验证功能，即资源服务器通过事先维护好的对称或非对称密钥（非对称密钥是授权服务器提供的公钥）直接在本地验证 token，这种去中心化的验证机制非常适合分布式架构。JWT 相对于传统的 token 来说，解决了以下两个"痛点"。

（1）通过验证签名，对 token 的验证可以直接在资源服务器本地完成，不需要连接授权服务器。

（2）在 Payload 中可以包含用户的相关信息，这样就轻松实现了 token 和用户信息的绑定。

如果授权服务器颁发的是 JWT 格式的 token，资源服务器就可以直接自己验证 token 的有效性并绑定用户信息，这无疑大大提高了处理效率且减少了单点隐患。

由于 Spring Boot 中的 OAuth 协议是在 Spring Security 的基础上完成的，因此通过 JWT 结合 Spring Security OAuth2 使用的方式，可以避免每次请求都需要远程调度认证授权服务。资源服务器只需要通过授权服务器验证一次，然后返回 JWT 即可。返回的 JWT 包含了用户的所有信息，包括权限信息。

7.1.3 在微服务架构中实现 SSO

SSO（Single Sign On，单点登录）可以为分布式环境中的不同应用提供统一的登录认证和授权管理。通过统一的登录认证和授权管理，用户只需要在任何应用中登录一次，就可以获得使用其他应用的权限。在用户微服务项目 user-microservice 中已经设计了权限管理业务，这里重点介绍认证和授权的实现。

为了方便认证，可以通过网关 Zuul 统一登录验证。同时，使用 RSA 算法生成公钥和私钥。私钥被保存在授权中心，公钥被保存在网关 Zuul 和各个微服务中。实现 SSO 的基本过程如图 7-2 所示。

（1）客户端（用户）发送请求，如果用户没有登录，则将进行登录，之后执行步骤 2；如果用户已经登录，则会携带 JWT，执行步骤 5。

（2）Zuul 请求授权。

（3）授权中心校验用户信息，在验证通过后，使用私钥生成 JWT，即使用私钥对 JWT 进行签名加密，并将 JWT 返回 Zuul。

Java EE 企业级应用开发技术研究

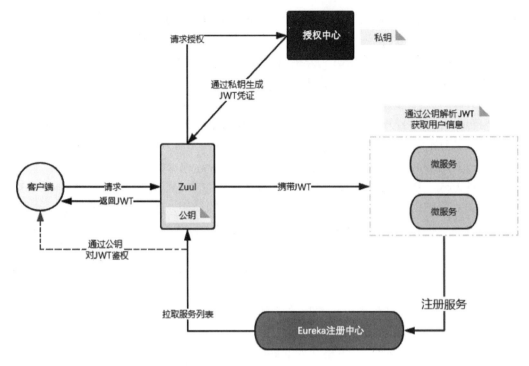

图 7-2 实现 SSO 的基本过程

（4）Zuul 返回 JWT 给客户端。

（5）Zuul 通过公钥对 JWT 鉴权，即直接通过公钥解密 JWT，并进行验证，若验证通过则放行。

（6）Zuul 从注册中心拉取服务列表，并根据用户的请求导航到客户端所访问的微服务。

（7）微服务直接使用公钥解析 JWT，获取用户信息。用户信息中包含用户的权限，根据用户的权限可以决定客户端是否有权访问相应的功能。

7.2 OAuth2 授权服务器模块设计

OAuth2 授权服务器可以为使用 SSO 的应用提供授权认证的服务。当用户登录一个应用后，就可以通过免登录的授权认证方式使用其他应用。

7.2.1 OAuth2 授权服务器模块

1. 创建 oauth-server 模块

在 base-microservice 项目中创建模块 oauth-server，并在 pom.xml 文件中引入如下依赖：

```
<dependency>
    <groupId>org.springframework.boot</groupId>
    <artifactId>spring-boot-starter-web</artifactId>
</dependency>
```

```xml
<dependency>
    <groupId>org.springframework.boot</groupId>
    <artifactId>spring-boot-devtools</artifactId>
    <scope>runtime</scope>
    <optional>true</optional>
</dependency>
<dependency>
    <groupId>org.springframework.cloud</groupId>
    <artifactId>spring-cloud-starter-netflix-eureka-client</artifactId>
</dependency>
<dependency>
    <groupId>org.springframework.cloud</groupId>
    <artifactId>spring-cloud-starter-oauth2</artifactId>
</dependency>
<dependency>
    <groupId>org.mybatis.spring.boot</groupId>
    <artifactId>mybatis-spring-boot-starter</artifactId>
    <version>2.1.0</version>
</dependency>
<dependency>
    <groupId>org.springframework.boot</groupId>
    <artifactId>spring-boot-starter-thymeleaf</artifactId>
</dependency>
<dependency>
    <groupId>com.fasterxml.jackson.core</groupId>
    <artifactId>jackson-core</artifactId>
    <version>2.10.3</version>
</dependency>
<dependency>
    <groupId>commons-beanutils</groupId>
    <artifactId>commons-beanutils</artifactId>
    <version>1.9.4</version>
</dependency>
<dependency>
    <groupId>com.alibaba</groupId>
    <artifactId>fastjson</artifactId>
    <version>1.2.47</version>
</dependency>
<dependency>
    <groupId>com.rc</groupId>
    <artifactId>user-clientapi</artifactId>
    <version>0.0.1-SNAPSHOT</version>
</dependency>
```

在上述依赖中，最关键的是 spring-cloud-starter-oauth2 依赖。此外，因为需要调用用户的 REST API 模块，所以需要引入 user-clientapi 依赖。

2．创建配置文件 application.properties

```
spring.application.name=oauth-server
eureka.client.service-url.defaultZone=http://localhost:8099/eureka/
```

```
server.port=5777
server.servlet.context-path=/oauth-server
spring.thymeleaf.cache=false
spring.thymeleaf.encoding=UTF-8
spring.thymeleaf.mode=HTML
spring.thymeleaf.prefix=classpath:/templates
spring.thymeleaf.suffix=.html
```

3．创建一个启动程序

```
package com.rc.oauth;
...//导包语句省略
@SpringBootApplication
@EnableDiscoveryClient
@EnableFeignClients(basePackages= {"com.rc.user.clientapi"})
public class OauthServerApplication {
    public static void main(String[] args) {
        SpringApplication.run(OauthServerApplication.class, args);
    }
}
```

7.2.2　对授权服务器进行配置

1．生成公钥和私钥

使用非对称密钥（公钥和私钥）执行签名过程，需要先生成公钥和私钥。

（1）生成用于 token 加密的私钥文件 rc-jwt.jks。jks 文件的生成需要使用 Java keytool 工具，保证 Java 环境变量无异常，输入命令如下：

```
keytool -genkeypair -alias rc-jwt
        -validity 3650
        -keyalg RSA
        -dname "CN=jwt,OU=jtw,O=jwt,L=zurich,S=zurich, C=CH"
        -keypass rc123456
        -keystore rc-jwt.jks
        -storepass rc123456
```

（2）导出公钥文件。资源微服务器需要使用 jks 文件的公钥对 JWT 进行解密。从 jks 文件中导出公钥的命令如下：

```
keytool -list -rfc -keystore rc-jwt.jks | openssl x509 -inform pem -pubkey
```

使用这个命令要求安装 OpenSSL，然后手动将安装的 openssl.exe 文件所在目录配置到环境变量中。

生成的公钥内容被存放在 key.cert 文件中，内容如下：

```
-----BEGIN PUBLIC KEY-----
MIGfMA0GCSqGSIb3DQEBAQUAA4GNADCBiQKBgQCErKFYow6Ab3OqTqJZAcbdnEIm
Pu/EUkjOWYFOY0CptnaAQwognziBAOEAlDqfnQ5Ona9huMr6aNks2Hlvq8+1gCUA
fC35bcQFft9pZIATscP5h+fYxB2KuqEflHxJVsimP2TDkUBIbfzUstPoYgriOMcC
+AoB5DMVanzL3nV6GwIDAQAB
-----END PUBLIC KEY-----
```

2. 定义 JWT 增强器

自定义一个 RcCustomTokenEnhancer 类。该类实现了 TokenEnhancer 接口，目的是使 JWT 包括权限信息，代码如下：

```java
package com.rc.oauth.config;
...//导包语句省略
public class RcCustomTokenEnhancer implements TokenEnhancer {
    @Override
    public OAuth2AccessToken enhance(OAuth2AccessToken accessToken, OAuth2Authentication authentication) {
        final Map<String, Object> additionalInfo = new HashMap<>();
        RcSysUser user = (RcSysUser) authentication.getUserAuthentication().getPrincipal();
        additionalInfo.put("authorities", user.getAuthorities());
        ((DefaultOAuth2AccessToken) accessToken).setAdditionalInformation(additionalInfo);
        return accessToken;
    }
}
```

3. 新建一个配置类 RcAuthorizationServerConfig

该配置类用于配置认证服务，代码如下：

```java
package com.rc.oauth.config;
...//导包语句省略
@Configuration
@EnableAuthorizationServer
public class RcAuthorizationServerConfig extends AuthorizationServerConfigurerAdapter {
    private static final String SCOPE = "all";
    private static final String CLIENT_ID = "zuul-server";
    private static final String CLIENT_SECRET = "123456";
    @Resource
    private AuthenticationManager authenticationManager;
    @Resource
    private UserDetailsService userDetailsService;
    @Resource
    private PasswordEncoder passwordEncoder;
    @Override
    public void configure(ClientDetailsServiceConfigurer clients) throws Exception {
        //配置客户端
        clients.inMemory().withClient(CLIENT_ID)
                .authorizedGrantTypes("password", "authorization_code", "refresh_token")
                .redirectUris("http://localhost:8084/login")
                .scopes(SCOPE)
                .autoApprove(true)
                .accessTokenValiditySeconds((int) TimeUnit.DAYS.toSeconds(1))
                .secret(passwordEncoder.encode(CLIENT_SECRET));
    }
    @Override
    public void configure(AuthorizationServerEndpointsConfigurer endpoints) throws Exception {
        //配置 JWT 增强器
        TokenEnhancerChain enhancerChain = new TokenEnhancerChain();
```

```
            enhancerChain.setTokenEnhancers(Arrays.asList(tokenEnhancer(), accessTokenConverter()));
            endpoints
                    .tokenStore(tokenStore())                           //自定义 token 生成方案
                    .tokenEnhancer(enhancerChain);
            endpoints
                    .authenticationManager(authenticationManager)   //身份认证管理器
                    .allowedTokenEndpointRequestMethods(HttpMethod.GET, HttpMethod.POST);
            endpoints
                    .userDetailsService(userDetailsService)
                    .reuseRefreshTokens(true);
        }
        @Override
        public void configure(AuthorizationServerSecurityConfigurer security) throws Exception {
            security
                    .allowFormAuthenticationForClients()
                    .tokenKeyAccess("permitAll()")
                    .checkTokenAccess("permitAll()")
                    .passwordEncoder(passwordEncoder);
        }
        @Bean
        public TokenStore tokenStore() {
            return new JwtTokenStore(accessTokenConverter());
        }
        @Bean
        public JwtAccessTokenConverter accessTokenConverter() {
            //配置 jks 文件
            JwtAccessTokenConverter converter = new JwtAccessTokenConverter();
            KeyPair keyPair = new KeyStoreKeyFactory(new ClassPathResource("rc-jwt.jks"),
"rc123456".toCharArray()).getKeyPair("rc-jwt ");
            converter.setKeyPair(keyPair);
            return converter;
        }
        @Bean
        public TokenEnhancer tokenEnhancer() {
            return new RcCustomTokenEnhancer();
        }
    }
```

在上述配置类中，使用了@EnableAuthorizationServer 标注来配置 OAuth2 授权服务机制，并使用了@Bean 标注的几个方法来一起配置这个授权服务机制。这个配置类继承于 AuthorizationServerConfigurerAdapter 类，主要通过 3 个 configure()方法进行配置。

（1）configure(ClientDetailsServiceConfigurer clients)方法：用来配置客户端详情服务（ClientDetailsService），即客户端详情信息会在这里进行初始化。我们可以将客户端详情信息写在这里，或者通过数据库来存储或调取详情信息。对于网上人才中心系统来说，客户端是 Zuul，相对简单，客户端详情信息可以直接写在代码配置中。

客户端详情信息主要涉及以下几方面。

- clientId：（必需的）用来标识客户的 ID。

- secret：（需要值得信任的客户端）客户端安全码。
- scope：用来限制客户端的访问范围。如果为空（默认），则表示客户端的访问范围没有任何限制。
- authorizedGrantTypes：客户端可以使用的授权类型。

（2）configure(AuthorizationServerEndpointsConfigurer endpoints)方法：用来配置授权（authorization）及令牌（token）的访问端点和令牌服务（token services）。

（3）configure(AuthorizationServerSecurityConfigurer security)方法：用来配置令牌端点（token endpoint）的安全约束。

除上述的 3 个方法外，下列方法主要用于配置 JWT。

（1）tokenStore()方法：返回一个 TokenStore 对象的子对象 JwtTokenStore。

（2）tokenEnhancer()方法：返回一个 token 增强器。

（3）jwtAccessTokenConverter()方法：返回一个 JwtAccessTokenConverter 对象。该对象用于根据签名生成 JwtToken。在这个方法中，根据密钥文件创建了 KeyPair 对象，并将该对象设置给 JwtAccessTokenConverter 对象。

7.2.3 登录管理及安全配置

1. 封装一个权限用户

```java
package com.rc.oauth.config;
...//导包语句省略
public class RcSysUser implements UserDetails {
    private RcUserQo user;
    public RcSysUser(RcUserQo user) {
        this.user = user;
    }
    @Override
    public Collection<? extends GrantedAuthority> getAuthorities() {
        List<SimpleGrantedAuthority> authorities = new ArrayList<>();
        //用于添加用户的权限。只要将用户权限添加到 authorities 中即可
        for (RcRoleQo role : user.getRoles()) {
            authorities.add(new SimpleGrantedAuthority(role.getRoleKey()));
            //System.out.println(role.getRoleName());
        }
        return authorities;
    }
    @Override
    public String getPassword() {
        return user.getUserPwd();
    }
    @Override
    public String getUsername() {
        return user.getUserName();
    }
    @Override
```

```java
    public boolean isAccountNonExpired() {
        return true;
    }
    @Override
    public boolean isAccountNonLocked() {
        return true;
    }
    @Override
    public boolean isCredentialsNonExpired() {
        return true;
    }
    @Override
    public boolean isEnabled() {
        return true;
    }
    public RcUserQo getUser() {
        return user;
    }
}
```

2. 定义一个用于登录验证的用户业务类

```java
package com.rc.oauth.config;
...//导包语句省略
@Service
public class YslntUserDetailsService implements UserDetailsService {
    @Resource
    private RcUserClientApi userClientApi;
    @Override
    public UserDetails loadUserByUsername(String username) throws UsernameNotFoundException {
        ModelMap map=userClientApi.findByUserName(username);
        if(map==null||map.get("code").toString().equals("0")){
            throw new UsernameNotFoundException("用户不存在!!!");
        }else {
            RcUserQo user=null;
            try{
                ObjectMapper objectMapper = new ObjectMapper();
                objectMapper.configure(DeserializationFeature.FAIL_ON_UNKNOWN_PROPERTIES, false);
                LinkedHashMap<String,Object> map1=(LinkedHashMap<String,Object>)map.get("data");
                user=objectMapper.readValue(objectMapper.writeValueAsString(map1),RcUserQo.class);
            }catch (Exception e){
            }
            if(user==null){
                throw new UsernameNotFoundException("用户不存在!!!");
            }
            return new YslntSysUser(user);
        }
    }
}
```

3. 安全配置

```java
package com.rc.oauth.config;
...//导包语句省略
@Configuration
@EnableWebSecurity
public class RcWebSecurityConfig extends WebSecurityConfigurerAdapter {
    @Resource
    private UserDetailsService userDetailsService;
    @Bean
    public AuthenticationFailureHandler authenctiationFailureHandler(){
        return new RcAuthenctiationFailureHandler();
    }
    @Bean
    public AuthenticationSuccessHandler authenctiationSuccessHandler(){
        return new RcAuthenctiationSuccessHandler();
    }
    @Bean
    @Override
    public AuthenticationManager authenticationManagerBean() throws Exception{
        return super.authenticationManagerBean();
    }
    @Override
    protected void configure(AuthenticationManagerBuilder auth) throws Exception {
        auth.userDetailsService(userDetailsService).passwordEncoder(passwordEncoder());
    }
    @Override
    public void configure(WebSecurity web) throws Exception {
        web.ignoring().antMatchers("/js/**", "/css/**", "/images/**");
    }
    @Override
    protected void configure(HttpSecurity http) throws Exception {
        http.cors().and().csrf().disable();
        http // 配置登录页/login 并允许访问
            .formLogin()
                .loginProcessingUrl("/signup")
                .loginPage("/login")
                .failureForwardUrl("/failure").permitAll()
            .and().logout().logoutUrl("/logout").logoutSuccessUrl("/")
            .and().authorizeRequests().mvcMatchers("/signup","/exit").permitAll()
            .anyRequest().authenticated();
    }
    @Bean
    public PasswordEncoder passwordEncoder() {
        return new BCryptPasswordEncoder();
    }
}
```

7.2.4 控制器和用户登录界面设计

1. 控制器设计

创建控制器类，代码如下：

```java
package com.rc.oauth.controller;
...//导包语句省略
@Controller
public class LoginController {
    @RequestMapping("/user")
    @ResponseBody
    public Principal user(Principal user){
        return user;
    }
    @GetMapping("/error")
    public String error() {
        return "/error";
    }
    @GetMapping("/login")
    public String login() {
        return "/login";
    }
    @GetMapping("/failure")
    public String failure() {
        return "/failure";
    }
    @RequestMapping("/exit")
    public String exit(HttpServletRequest request, HttpServletResponse response) {
        try {
            new SecurityContextLogoutHandler().logout(request, null, null);
            return "/logout";
        } catch (Exception e) {
            e.printStackTrace();
            return "/error";
        }
    }
}
```

2. 用户登录界面设计

在 resources/template 文件夹下建立用户登录界面 login.html，其运行效果如图 7-3 所示。

第 7 章 在微服务架构中整合 OAuth2

图 7-3 用户登录界面的运行效果

示例代码如下:

```
<!DOCTYPE html>
<html xmlns:th="http://www.thymeleaf.org">
<head>
    <meta http-equiv="Content-Type" content="text/html; charset=UTF-8"/>
    <title>网上人才中心系统</title>
    <link rel="stylesheet" th:href="@{/css/all.css}" type="text/css"/>
    <script>
        if (top.location !== self.location) {
            top.location.href = self.location.href;
        }
    </script>
</head>
<body>
<div id="container">
    <div id="header"></div>
    <div id="pagebody" class="back">
        <form id="loginform" method="post" th:action="@{/signup}">
            <p><span>用户名</span>
            <input type="text" name="username" size="20" value></p>
                <p><span>密　码</span>
            <input type="password" name="password" size="20" value></p>
                <p> <input class="button ok" type="submit" value=" 登 录 ">
                    <input class="button ok" type="button" value=" 注 册 ">
```

```
            </p>
        </form>
    </div>
    <div id="footer">
        <hr/>
        Copyright &copy; 2020-2030 YSL.All Rights Reserved.
    </div>
</div>
</body>
</html>
```

7.3 实现微服务应用访问控制

7.3.1 对网关 Zuul 进行配置

在通过网关 Zuul 进行统一登录验证时，需要对网关 Zuul 进行配置。

1．引入依赖

在 zuul-server 模块的 pom 文件中引入如下依赖：

```
<dependency>
    <groupId>org.springframework.cloud</groupId>
    <artifactId>spring-cloud-starter-oauth2</artifactId>
</dependency>
```

2．增加安全配置

在 zuul-server 模块中，增加如下安全配置：

```
package com.rc.gateway.config;
...//导包语句省略
@EnableOAuth2Sso
@Configuration
public class WebSecurityConfig extends WebSecurityConfigurerAdapter {
    @Override
    protected void configure(HttpSecurity http) throws Exception {
        http.logout().logoutSuccessUrl("http://localhost:5777/oauth-server/exit");
        http.authorizeRequests()
                .antMatchers(
                        "/swagger-ui.html",
                        "/swagger-resources/**",
                        "/v2/api-docs",
                        "/webjars/**"
                ).permitAll()
                .anyRequest().authenticated();
        http.csrf().disable();
        http.headers().frameOptions().disable();
    }
}
```

第 7 章　在微服务架构中整合 OAuth2

3．修改启动程序

在启动程序中增加@EnableZuulProxy 配置，以便传递 token：

```
package com.rc.gateway;
...//导包语句省略
@SpringBootApplication
@EnableZuulProxy
@EnableDiscoveryClient
public class GatewayServerApplication {
    public static void main(String[] args) {
        f.run(GatewayServerApplication.class, args);
    }
}
```

4．bootstrap.properties 文件

修改 bootstrap.properties 文件，增加如下安全配置：

```
security.oauth2.client.client-id=gateway-server
security.oauth2.client.client-secret=123456
security.oauth2.client.grant-type=authorization_code
security.oauth2.client.scope=all
security.oauth2.client.access-token-uri=http://localhost:5777/oauth-server/oauth/token
security.oauth2.client.user-authorization-uri=http://localhost:5777/oauth-server/oauth/authorize
#非对称，通过访问/oauth/token_key 获得公钥并进行解析
security.oauth2.resource.jwt.key-uri=http://localhost:5777/oauth-server/oauth/token_key
```

7.3.2　创建安全模块

在微服务架构中，由于一个微服务应用相当于一个资源服务器，各资源服务器具有类似的配置，因此，我们可以先创建一个安全模块。

1．创建 base-security 模块

在 base-microservice 项目中创建 base-security 模块，引入如下依赖：

```xml
<dependency>
    <groupId>org.springframework.boot</groupId>
    <artifactId>spring-boot-starter-web</artifactId>
</dependency>
<dependency>
    <groupId>org.springframework.cloud</groupId>
    <artifactId>spring-cloud-starter-oauth2</artifactId>
</dependency>
<dependency>
    <groupId>com.rc</groupId>
    <artifactId>user-clientapi</artifactId>
    <version>0.1-SNAPSHOT</version>
</dependency>
```

2．创建权限验证失败处理类

```
package com.rc.security;
```

```
...//导包语句省略
public class RcCustomAccessDeniedHandler implements AccessDeniedHandler {
    @Override
    public void handle(HttpServletRequest request, HttpServletResponse response,
AccessDeniedException e) throws IOException, ServletException {
        response.setContentType("text/html;charset=UTF-8");
        String baseUrl=request.getServletPath();
        StringBuffer sb=new StringBuffer();
        sb.append("<!DOCTYPE html>");
        sb.append("<html>");
        sb.append("<head>");
        sb.append("<meta http-equiv=\"Content-Type\" content=\"text/html; charset=UTF-8\"/>");
        sb.append("<title>网上人才中心系统</title>");
        sb.append("<link rel=\"stylesheet\" href=\"http://localhost:5777/oauth-server/css/all.css\" type=\"text/css\"/>");
        sb.append("<script>");
        sb.append("if (top.location !== self.location) {");
        sb.append("    top.location.href = self.location.href;");
        sb.append("}");
        sb.append("</script>");
        sb.append("</head>");
        sb.append("<div id=\"container\">");
        sb.append("<div id=\"header\"></div>");
        sb.append("<div id=\"pagebody\" style=\"text-align:center;padding-top:50px;\">");
        sb.append("权限不足！");
        sb.append("</div>");
        sb.append("<div id=\"footer\"><hr/>");
        sb.append("Copyright &copy; 2020-2030 YSL.All Rights Reserved.");
        sb.append("</div>");
        sb.append("</div>");
        sb.append("</body>");
        sb.append("</html>");
        response.setStatus(HttpServletResponse.SC_UNAUTHORIZED);
        response.getWriter().write(sb.toString());
    }
}
```

3. 创建异常处理类

```
package com.rc.security;
...//导包语句省略
public class RcAuthExceptionEntryPoint implements AuthenticationEntryPoint {
    @Override
    public void commence(HttpServletRequest request, HttpServletResponse response,
AuthenticationException authException) throws IOException, ServletException {
        String msg;
        Throwable cause = authException.getCause();
        if(cause instanceof InvalidTokenException) {
            msg="无效的token";
        }else{
```

```java
            msg="访问此资源需要完全的身份验证";
        }
        String baseUrl=request.getServletPath();
        response.setContentType("text/html;charset=UTF-8");
        StringBuffer sb=new StringBuffer();
        sb.append("<!DOCTYPE html>");
        sb.append("<html>");
        sb.append("<head>");
        sb.append("<meta http-equiv=\"Content-Type\" content=\"text/html; charset=UTF-8\"/>");
        sb.append("<title>网上人才中心系统</title>");
        sb.append("<link rel=\"stylesheet\" href=\"http://localhost:5777/oauth-server/css/all.css\" type=\"text/css\"/>");
        sb.append("<script>");
        sb.append("if (top.location !== self.location) {");
        sb.append("    top.location.href = self.location.href;");
        sb.append(")");
        sb.append("</script>");
        sb.append("</head>");
        sb.append("<div id=\"container\">");
        sb.append("<div id=\"header\"></div>");
        sb.append("<div id=\"pagebody\" style=\"text-align:center;padding-top:50px;\">");
        sb.append(msg);
        sb.append("</div>");
        sb.append("<div id=\"footer\"><hr/>");
        sb.append("Copyright &copy; 2020-2030 YSL.All Rights Reserved.");
        sb.append("</div>");
        sb.append("</div>");
        sb.append("</body>");
        sb.append("</html>");
        response.setStatus(HttpServletResponse.SC_UNAUTHORIZED);
        response.getWriter().write(sb.toString());
    }
}
```

4．自定义 AccessTokenConverter 转换类

定义此类的主要目的是从 token 中获取权限信息，代码如下：

```java
package com.rc.security;
...//导包语句省略
@Component
public class RcCustomerAccessTokenConverter extends DefaultAccessTokenConverter {

    public RcCustomerAccessTokenConverter() {
        super.setUserTokenConverter(new RcCustomerUserAuthenticationConverter());
    }
    private class RcCustomerUserAuthenticationConverter extends DefaultUserAuthenticationConverter {
        String AUTHORITIES="authorities";
        @Override
        public Authentication extractAuthentication(Map<String, ?> map) {
            return new UsernamePasswordAuthenticationToken(map, "N/A", this.getAuthorities(map));
```

```java
        }
        private Collection<? extends GrantedAuthority> getAuthorities(Map<String, ?> map) {
            if (!map.containsKey(AUTHORITIES)) {
                return AuthorityUtils.commaSeparatedStringToAuthorityList(StringUtils
                    .arrayToCommaDelimitedString(new String[]{}));
            }else {
                Object authorities = map.get(AUTHORITIES);
                if (authorities instanceof String) {
                    return AuthorityUtils.commaSeparatedStringToAuthorityList((String) authorities);
                } else if (authorities instanceof Collection) {
                    List<GrantedAuthority> list=AuthorityUtils.commaSeparatedStringToAuthorityList(
                    StringUtils.collectionToCommaDelimitedString((Collection<?>) authorities));
                    List<GrantedAuthority> list1=new ArrayList<>();
                    for(GrantedAuthority ga:list) {
                        try {
                            String role=ga.getAuthority();
                            role=role.substring(role.indexOf("=")+1,role.length()-1);
                            list1.add(new SimpleGrantedAuthority(role));
                        }catch(Exception e){
                        }
                    }
                    return list1;
                } else {
                    throw new IllegalArgumentException("Authorities must be either a String or a Collection");
                }
            }
        }
    }
}
```

5. 定义决策管理器

```java
package com.rc.security;
...//导包语句省略
@Component
@Slf4j
public class RcAccessDecisionManager implements AccessDecisionManager {
    @Override
    public void decide(Authentication authentication, Object object, Collection<ConfigAttribute> configAttributes)
            throws AccessDeniedException, InsufficientAuthenticationException {
        if (null == configAttributes || configAttributes.size() <= 0) {
            return;
        }
        ConfigAttribute c;
        String needRole;
        for (Iterator<ConfigAttribute> iter = configAttributes.iterator(); iter.hasNext();) {
            c = iter.next();
            needRole = c.getAttribute();
```

```java
            for (GrantedAuthority ga : authentication.getAuthorities()) {
                if (needRole.trim().equals(ga.getAuthority())) {
                    return;                                    //有权限放行
                }
            }
        }
        log.debug(authentication.toString());
        throw new AccessDeniedException("access denied");       //无权限返回
    }

    @Override
    public boolean supports(ConfigAttribute attribute) {
        return true;
    }
    @Override
    public boolean supports(Class<?> clazz) {
        return true;
    }
}
```

6．创建安全源数据业务类

```java
package com.rc.security;
...//导包语句省略
@Component
public class RcInvocationSecurityMetadataSourceService implements
                FilterInvocationSecurityMetadataSource, InitializingBean {
    @Resource
    private RcPermissionClientApi rcPermissionClientApi;
    private static ConcurrentHashMap<String,Collection<ConfigAttribute>> map = new 
ConcurrentHashMap<String, Collection<ConfigAttribute>>();
    public void loadResourceDefine() {
        ModelMap map1 = rcPermissionClientApi.findAll();         //获取所有配置权限
        Collection<ConfigAttribute> array;
        ConfigAttribute cfg;
        map.clear();
        //map = new HashMap<>();
        for (String url : map1.keySet()) {
            array = new ArrayList<>();
            for(String roleKey:(ArrayList<String>)map1.get(url)){
                array.add(new SecurityConfig(roleKey));
            }
            //使用权限的 getUrl()方法返回值作为 map 的 key，使用 ConfigAttribute 的集合作为 value
            map.put(url, array);
        }
    }
    @Override
    public Collection<ConfigAttribute> getAttributes(Object object) throws IllegalArgumentException {
        loadResourceDefine();
        HttpServletRequest request = ((FilterInvocation) object).getHttpRequest();
```

```java
            AntPathRequestMatcher matcher;
            String resUrl;
            for (Iterator<String> iter = map.keySet().iterator(); iter.hasNext();) {
                resUrl = iter.next();
                matcher = new AntPathRequestMatcher(resUrl);
                if (matcher.matches(request)) { return map.get(resUrl); }
            }
            return null;
    }
    @Override
    public Collection<ConfigAttribute> getAllConfigAttributes() {
        return null;
    }
    @Override
    public boolean supports(Class<?> clazz) {
        return true;
    }
    @Override
    public void afterPropertiesSet() throws Exception {
        loadResourceDefine();
    }
}
```

7. 创建权限拦截器类

```java
package com.rc.security;
...//导包语句省略
@Component
public class RcFilterSecurityInterceptor extends AbstractSecurityInterceptor implements Filter {
    @Resource
     private RcInvocationSecurityMetadataSourceService rcInvocationSecurityMetadataSourceService;
    @Resource
    @Override
    public void setAccessDecisionManager(AccessDecisionManager rcAccessDecisionManager) {
        super.setAccessDecisionManager(rcAccessDecisionManager);
    }
    @Override
    public void init(FilterConfig filterConfig) throws ServletException {

    }
    @Override
    public void doFilter(ServletRequest request, ServletResponse response, FilterChain chain)
            throws IOException, ServletException {
        FilterInvocation fi = new FilterInvocation(request, response, chain);
        invoke(fi);
    }
    public void invoke(FilterInvocation fi) throws IOException, ServletException {
        InterceptorStatusToken token = super.beforeInvocation(fi);
        try {
            //执行下一个拦截器
```

```
                fi.getChain().doFilter(fi.getRequest(), fi.getResponse());
            } finally {
                super.afterInvocation(token, null);
            }
    }
    @Override
    public void destroy() {}
    @Override
    public Class<?> getSecureObjectClass() {
        return FilterInvocation.class;
    }
    @Override
    public SecurityMetadataSource obtainSecurityMetadataSource() {
        return this.rcInvocationSecurityMetadataSourceService;
    }
}
```

8. 配置资源服务器

当采用非对称加密方式时,资源服务器会通过公钥来验证 token,因此需要将公钥文件 key.cert 存放在 resources 目录下。

创建一个 RcResourceServerConfig 类,用来继承 ResourceServerConfigurerAdapter 类,并使用@EnableResourceServer 标注,代码如下:

```
package com.rc.security;
...//导包语句省略
@Configuration
@EnableResourceServer
public class RcResourceServerConfig extends ResourceServerConfigurerAdapter {
    @Resource
    RcFilterSecurityInterceptor rcFilterSecurityInterceptor;    //使用自定义的方法放行判定类
    @Override
    public void configure(HttpSecurity http) throws Exception {
        http.headers().frameOptions().disable();
        http
                .csrf().disable()
                .logout()
                .logoutSuccessUrl("http://localhost:8084/logout")
                .and().authorizeRequests()
                .antMatchers(
                        "/swagger-ui.html",              //让它们可以通过 Spring 的加密体系
                        "/swagger-resources/**",
                        "/v2/api-docs",
                        "/webjars/**"
                ).permitAll()
                .anyRequest().authenticated();
        //添加自定义方法的过滤器
        http.addFilterBefore(rcFilterSecurityInterceptor, FilterSecurityInterceptor.class);
    }
    @Bean
```

```java
    public JwtAccessTokenConverter jwtAccessTokenConverter() throws IOException {
        JwtAccessTokenConverter jwtAccessTokenConverter = new JwtAccessTokenConverter();
        ClassPathResource resource = new ClassPathResource("key.cert");
        jwtAccessTokenConverter.setVerifierKey(new String(FileCopyUtils.copyToByteArray(resource.getInputStream())));
        jwtAccessTokenConverter.setAccessTokenConverter(new RcCustomerAccessTokenConverter());
        return jwtAccessTokenConverter;
    }
    @Bean
    public TokenStore tokenStore() throws IOException {
        return new JwtTokenStore(jwtAccessTokenConverter());
    }
    @Override
    public void configure(ResourceServerSecurityConfigurer resources) throws Exception {
        resources.tokenStore(tokenStore()).authenticationEntryPoint(new AuthExceptionEntryPoint())
        .accessDeniedHandler(new RcCustomAccessDeniedHandler());;
    }
}
```

7.3.3 配置微服务应用

1. 引入依赖

在需要进行访问控制的微服务中，引入如下依赖：

```xml
<dependency>
    <groupId>com.rc</groupId>
    <artifactId>base-security</artifactId>
    <version>0.1-SNAPSHOT</version>
</dependency>
```

2. 修改启动类

对启动程序进行修改，增加包的指定配置，以便快速找到所需的包。以 company-web 模块为例，配置代码如下：

```java
package com.rc.company;
...//导包语句省略
@SpringBootApplication(exclude = DataSourceAutoConfiguration.class,scanBasePackages = {"com.rc.company","com.rc.security"})
@EnableDiscoveryClient
@EnableFeignClients(basePackages = {"com.rc.user.clientapi","com.rc.company.clientapi"})
public class CompanyWebApplication {
    public static void main(String[] args) {
        SpringApplication.run(CompanyWebApplication.class, args);
    }
}
```

参考文献

[1] 企业级应用的概念和特点[EB/OL]. [2006-07-28]. https://blog.csdn.net/sorichwalk/article/details/994097.

[2] MyBatis[EB/OL]. [2010-02-01]. https://baike.baidu.com/item/MyBatis/2824918?fr=aladdin.

[3] 杨树林，胡洁萍. Java EE 企业级架构开发技术与案例教程[M]. 北京：机械工业出版社，2011.

[4] PlantUML 入门[EB/OL]. [2019-03-24]. https://www.jianshu.com/p/4068e5cf8355.

[5] 使用 Swagger2 自动生成 API 接口文档[EB/OL]. [2019-12-18]. https://zhuanlan.zhihu.com/p/98075551.

[6] 陈韶健. 深入实践 Spring Boot[M]. 北京：机械工业出版社，2016.

[7] Spring Boot 基础之 MockMvc 单元测试[EB/OL]. [2019-04-02]. https://blog.csdn.net/wo541075754/article/details/88983708.

[8] Spring Data JPA-Reference Documentation[EB/OL]. [2017-03-02]. https://docs.spring.io/spring-data/jpa/docs/1.11.1.RELEASE/reference/html/#jpa.query.spel-expressions.

[9] Spring Boot 数据源配置及数据库连接池的概念[EB/OL]. [2019-05-08]. http://www.west999.com/info/html/chengxusheji/Javajishu/20190508/4642202.html.

[10] 数据库连接池技术之 HikariCP[EB/OL]. [2020-02-02]. https://blog.csdn.net/lianghecai52171314/article/details/104143621.

[11] SpringBoot 中 Druid 数据源配置[EB/OL]. [2019-02-18]. https://blog.csdn.net/qq_41320105/article/details/87636227.

[12] 杨开振. 深入浅出 MyBatis 技术原理与实战[M]. 北京：电子工业出版社，2016.

[13] MyBatis Generator[EB/OL]. [2019-11-24]. http://www.mybatis.org/generator/generatedobjects/exampleClassUsage.html.

[14] MyBatis-Plus[EB/OL]. [2020-03-25]. https://mybatis.plus/guide/generator.html.

[15] Spring Security 工作原理概览[EB/OL]. [2019-04-27]. https://blog.csdn.net/u012702547/article/details/89629415.

[16] SpringBoot 整合 Spring Security[EB/OL]. [2019-09-04]. https://blog.csdn.net/duyunzhi/article/details/100532494.

[17] SpringSecurity 入门-SpringBoot 集成 SpringSecurity[EB/OL]. [2019-11-26]. https://www.itsource.cn/web/news/2233.html.

[18] Micro-service-learning. [EB/OL]. [2019-05-16]. https://github.com/taoweidong/Micro-service-learning.

[19] 微服务架构最强详解[EB/OL]. [2019-07-01]. https://www.cnblogs.com/kenshinobiy/p/11113124.html.

[20] 陈韶健. Spring Cloud 与 Docker 高并发微服务架构设计实施[M]. 北京：电子工业出版社，2108.

[21] SpringCloud 微服务学习笔记[EB/OL]. [2019-06-10]. https://blog.csdn.net/taoweidong1/article/details/91405599.

[22] Spring Cloud 微服务架构学习笔记与示例[EB/OL]. [2018-09-12]. https://www.cnblogs.com/edisonchou/p/java_spring_cloud_foundation_sample_list.html.

[23] SpringBoot-使用 Spring Security 实现 OAuth2 授权认证教程（实现 token 认证）[EB/OL]. [2020-01-08]. https://www.hangge.com/blog/cache/detail_2683.html.

[24] Spring Security 与 OAuth2（介绍）[EB/OL]. [2018.01.23]. https://www.jianshu.com/p/68f22f9a00ee.

反侵权盗版声明

电子工业出版社依法对本作品享有专有出版权。任何未经权利人书面许可,复制、销售或通过信息网络传播本作品的行为;歪曲、篡改、剽窃本作品的行为,均违反《中华人民共和国著作权法》,其行为人应承担相应的民事责任和行政责任,构成犯罪的,将被依法追究刑事责任。

为了维护市场秩序,保护权利人的合法权益,我社将依法查处和打击侵权盗版的单位和个人。欢迎社会各界人士积极举报侵权盗版行为,本社将奖励举报有功人员,并保证举报人的信息不被泄露。

举报电话:(010)88254396;(010)88258888
传　　真:(010)88254397
E-mail:dbqq@phei.com.cn
通信地址:北京市万寿路 173 信箱
　　　　　电子工业出版社总编办公室
邮　　编:100036